# ENCYCLOPAEDIA OF WORLD GEOGRAPHY

# ENCYCLOPAEDIA OF WORLD GEOGRAPHY

*By*

**R.V. Prajapati**

**2008**

**CYBER TECH PUBLICATIONS**
4264/3, Ansari Road, Daryaganj, New Delhi-110002 (India)
Ph.: 011-23244078. Fax: 011-23280028
E-Mail: cyberpublicationsdelhi@yahoo.com

ENCYCLOPAEDIA OF WORLD GEOGRAPHY

R.V. Prajapati



First Edition 2008



*Published by :*

**G.S. Rawat** for **Cyber Tech Publications**
4264/3, Ansari Road, Daryaganj, New Delhi-110002 (India)
Ph.: 011-23244078. Fax: 011-23280028
E-Mail: cyberpublicationsdelhi@yahoo.com

*Printed at :* **Amit Enterprises**, Maujpur, Delhi, (9810878184)

# Preface

The study of the natural environment, geography deals with the diverse physical, biological, and cultural features of the Earth's surface and examines how humans interact with it.. World geography is an important topic for everyone. It can be used to study the past but is also crucial to understanding the events and current political conditions of the modern world.

Encyclopedia of World Geography is an authoritative book including everything from the Fulda Gap to the Hudson River. Approximately 750 entries describe pioneers in geography and mapmaking; regions, boundaries, and places; and tools and terminology of geography and cartography. This accessible resource includes a lot of necessary information regarding the climate, major places, geographical conditions etc of the whole world. Some 100 scholars and experts in the field contributed entries to the text, and a senior academic served as editor. An invaluable reference, Encyclopedia of World Geography is a complete and comprehensive guide to the history and practice of geography and cartography.

However it will be exaggeration to say that I have did a much to bring this book in this form, I have heavily drawn upon from a numerous sources and books of reputed authors. I owe my debt and gratitude to all of them. Although every possible care has been taken to eliminate mistakes but still some are likely to remain in this realm. Readers are kindly requested to give their feedbacks as we can improve this book in its forthcoming editions.

—*Author*

# Contents

# 1

# Basic Facts of Geography

Geography is the study of the earth and its features, inhabitants, and phenomena. A literal translation would be "to describe the Earth". The first person to use the word "geography" was Eratosthenes (275-195 B.C.). Four historical traditions in geographical research are the spatial analysis of natural and human phenomena (geography as a study of distribution), area studies (places and regions), study of man-land relationship, and research in earth sciences. Nonetheless, modern geography is an all-encompassing discipline that foremost seeks to understand the world and all of its human and natural complexities-- not merely where objects are, but how they have changed and come to be. As "the bridge between the human and physical sciences," geography is divided into two main branches - human geography and physical geography.

## INTRODUCTION

Traditionally, geography as well as geographers has been viewed as the same as cartography and people who study place names. Although many geographers are trained in toponymy and cartography, this is not their main preoccupation. Geographers study the spatial and temporal distribution of phenomena, processes and feature as well as the interaction of humans and their environment. As space and place affect a variety of topics such as economics, health, climate, plants and animals, geography is highly interdisciplinary.

*"mere names of places...are not geography...To know by heart a whole gazetteer full of them would not, in itself, constitute anyone a geographer. Geography has higher*

*aims than this: it seeks to classify phenomena (alike of the natural and of the political world, in so far as it treats of the latter), to compare, to generalize, to ascend from effects to causes, and, in doing so, to trace out the great laws of nature and to mark their influences upon man. This is 'a description of the world'-that is Geography. In a word Geography is a Science-a thing not of mere names but of argument and reason, of cause and effect."* - *William Hughes 1863*

Geography as a discipline can be split broadly into two main sub fields: human geography and physical geography. The former focuses largely on the built environment and how space is created, viewed and managed by humans as well as the influence humans have on the space they occupy. The latter examines the natural environment and how the climate, vegetation & life, soil, water and landforms are produced and interact. As a result of the two subfields using different approaches a third field has emerged, which is environmental geography. Environmental geography combines physical and human geography and looks at the interactions between the environment and humans.

## HISTORY OF GEOGRAPHY

The ideas of Anaximander of Miletus (c. 610 B.C.-c. 545 B.C.), considered by later Greek writers to be the true founder of geography, come to us through fragments quoted by his successors. Anaximander is credited with the invention of the gnomon, the simple yet efficient Greek instrument that allowed the early measurement of latitude. Thales, Anximander is also credited with the prediction of eclipses. The foundations of geography can be traced to the ancient cultures, such as the ancient, medieval, and early modern Chinese. The Greeks, who were the first to explore geography as both art and science, achieved this through Cartography, Philosophy, and Literature, or through Mathematics. There is some debate about who was the first person to assert that the Earth is spherical in shape, with the credit going either to Parmenides or Pythagoras. Anaxagoras was able to demonstrate that the profile of the

Earth was circular by explaining eclipses. However, he still believed that the Earth was a flat disk, as did many of his contemporaries. One of the first estimates of the radius of the Earth was made by Eratosthenes.

The first rigorous system of latitude and longitude lines is credited to Hipparchus. He employed a sexagesimal system that was derived from Babylonian mathematics. The parallels and meridians were sub-divided into 360°, with each degree further subdivided 60' (minutes). To measure the longitude at different location on Earth, he suggested using eclipses to determine the relative difference in time. The extensive mapping by the Romans as they explored new lands would later provide a high level of information for Ptolemy to construct detailed atlases. He extended the work of Hipparchus, using a grid system on his maps and adopting a length of 56.5 miles for a degree.

During the Middle Ages, the fall of the Roman empire led to a shift in the evolution of geography from Europe to the Islamic world. Scholars such as Idrisi (produced detailed maps), Ibn Batutta, and Ibn Khaldun provided detailed accounts of their Hajj. Further, Islamic scholars translated and interpreted the earlier works of the Romans and Greeks and established the House of Wisdom in Baghdad for this purpose. From the 3rd century onwards, Chinese methods of geographical study and writing of geographical literature became much more complex than what was found in Europe at the time (until the 13th century). Chinese geographers such as Liu An, Shen Kuo, Fan Chengda, Zhou Daguan, and Xu Xiake wrote important treatises, yet by the 17th century, advanced ideas and methods of Western-style geography were adopted in China.

The Age of discovery during the 16th and 17th centuries where many new lands were discovered and accounts by explorers such as Christopher Columbus, Marco Polo and James Cook, revived a desire for both accurate geographic detail, and more solid theoretical foundations.

The 18th and 19th centuries were the times when geography became recognized as a discrete academic discipline and became part of a typical university curriculum in Europe (especially Paris and Berlin). The development of many geographic societies also occurred during the 19th century with the foundations of the Société de Géographie in 1821, the Royal Geographical Society in 1830, Russian Geographical Society in 1845, American Geographical Society in 1851, and the National Geographic Society in1888. The influence of Immanuel Kant, Alexander von Humbolt, Carl Ritter and Paul Vidal de la Blache can be seen as a major turning point in geography from a philosophy to an academic subject.

Over the past two centuries the advancements in technology such as computers, have led to the development of geomatics and new practices such as participant observation and geostatistics being incorporated into geography's portfolio of tools. In the West during the 20th century, the discipline of geography went through four major phases: environmental determinism, regional geography, the quantitative revolution, and critical geography. The strong interdisciplinary links between geography and the sciences of geology and botany, as well as economics, sociology and demographics have also grown greatly especially as a result of Earth System Science that seeks to understand the world in a holistic view.

## AGE OF GLOBALIZATION

Globalization: what is it and what does it mean to me? Economically? Technologically? Politically? Socially? Morally?

We constantly hear about how the world is getting smaller--is it? No, the world is becoming more connected. More information is flowing faster than ever before. For the first time EVER any of us could travel to ANY part of the world virtually overnight. We can communicate in REAL TIME with any part of the globe.

We are the first generation of humans, in all human history to enjoy foreign travel as a casual part of life, to communicate by direct-dialing with any country on all continents, to receive instant news of world happenings, to be expected to work overseas or work for a company that deals overseas: EXPECTED, NOT THE EXCEPTION. This is a really important concept, especially to you--the first generation that is living in the post-industrial, high technology, interconnected age.

Many if not all of you will work for multinational companies whose business is all over the world. Many if not all of you will work and live outside of the US at some point in your careers. People who understand and can interpret the world around them are much more versatile and therefore quickly become NON-EXPENDABLE. Businesses and jobs are internationalizing as we speak-almost all jobs, not just the fancy ones. You guys are the people that are going to be running the world. You guys are the decision makers--when all is said and done I'm just a dude teaching here at Virginia Tech. But YOU all will be the ones building the bridges, and electing leaders, and stabilizing governments, and controlling monetary exchange rates, or setting up all sorts of private or national or even international businesses/programs/projects that will shape the world and all the people on it.

Make no mistake about it, the AIDS rate in Africa DOES affect you, the increasing coal consumption in China DOES affect you, an earthquake in Japan, and the price of cocaine leaves in Colombia DOES affect you. Globalization is pretty much a one-way street. We are not going back to medieval times, no matter what isolationist say, do, or think. Ignore the rest of the world at your own peril-you wont be hurting anyone but yourself.

Knowledge is power, or at least empowerment. The more you know about the world in which we live, the more power you have.

## WHAT IS GEOGRAPHY?

Geography is one of the most fundamental sciences, a discipline awakened and informed by a long-standing human curiosity about our surroundings and the world environment. The term geography has its roots in the Greek words for "describing the earth", and this discipline has been carried forward since classical times by all cultures and civilizations. With the inherent satisfaction about knowing different environments comes pragmatic needs that compliment exploration, resource exploitation, world commerce, and travel. In some ways, geography can be compared to history: while that field describes and explains what has happened over time, geography describes and explains Earth's spatial dimension--of how the world differs from place to place. However, geography is a very holistic field; given that to describe any place on earth fully, we must include its history, its physical properties, its people, etc. etc.--so much so, that a myriad of different fields could be examined in geographic, or spatial, terms for any place on earth.

Given the broad scope of the geographical charge, it is no surprise that geographers have many different approaches to studying the world based upon conceptual emphasis guiding their study. At the most basic level, geography can be broken into two complementary pursuits, physical and human geography. As the term suggests, physical geography examines things like climate, landforms, soils, vegetation, and animal life, while human--or cultural--geography concentrates on the spatial analysis of economic, political, religious, architecture or farming systems. For example, the physical geographer in the Amazon river basin might be interested primarily in the ecological diversity of the tropical rain forest or how the destruction of that dense vegetation changes the local climate or hydrology. The human geographer, on the other hand, might focus on the social and economic factors explaining migration of settlers into the rain forest or the tensions and conflicts over land and resources between these people and indigenous peoples.

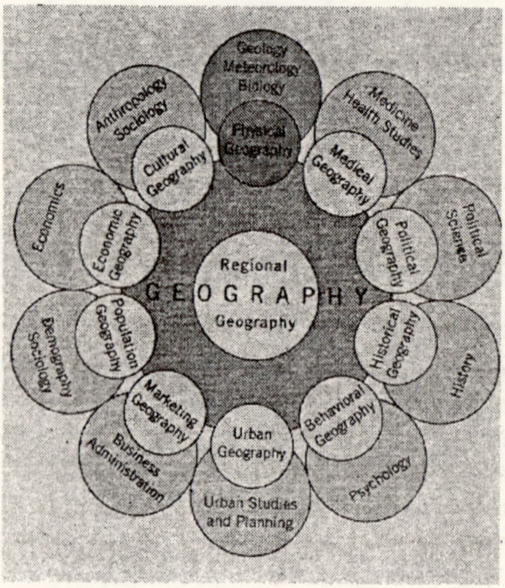

## APPROACHES TO GEOGRAPHY

All geographers collect and analyze geographic information, but they focus on different topics and employ different analytical approaches. What I love about geography is the fact that we can study virtually ANYTHING on earth in a geographic fashion--since everything in our lives occurs on earth by default we can study it in the context of where it happens, and ask other questions like why, when and how. Here are three basic approaches to consider when studying anything geographically:

- As a geographer, you can study Earth in a context of interrelated environmental (physical) and human systems. Systems are interdependent groups of items that interact in a regular way to form a unified whole. We can focus on physical systems such as climate or volcanoes, on human systems such as religions or economies, or even human-environmental relationships such as how a places resources or terrain (physical) affect its local economy (cultural). We have whole branches of specialization within this context that are tied to other independent fields--you may study how weather works in meteorology, but if you study weather on a spatial context (where certain patterns of rainfall occur, how temperature affects vegetation in an area, etc.) then you would be in the geographic specialization of climatology. The diagram to the right outlines a few other such relationships in this systematic approach in geography.

- We can also simply take a spatial analysis approach to any particular subject--be it a person, place or thing, or even a process. Thinking spatially means asking the question 'where' about a phenomenon, and describing or developing your answer around this question. After determining 'where', you might then ask other questions like 'why here?' 'why at this time?', 'how did it get there, or how did it develop?'. You can use this approach to literally study anything on earth--the distribution of cockroaches, the travel path of Alexander the Great, the diffusion of rock and roll music, the relationship between tornado paths and trailer parks, market distribution of

Coors beer, fan distribution of the Chicago Bulls--you name it, we can study it geographically.

- Finally, and most pertinent to our class, we can use area analysis which integrates the geographic features of an area or place in a method of study. Geographers have a long tradition (3 or 4 thousand years) of surveying, describing, and compiling geographic data about places. Each place in the world occupies a unique location and features a unique combination of human behavior and environmental processes that give it a special character. For purposes of geographic study, we divide the earth into regions, which are areas defined by one or more distinctive characteristic or features, such as climate, soil type, language, or economic activity. Certain human activities and environmental processes give their regions unified character and distinguish them from other areas of earth's surface. We can define our regions strictly by physical characteristics such as climate zone, strictly by human characteristics such as political entities (i.e. Mexico is a region defined by its political borders), or a combination of physical and human characteristics such as the region we call Eastern Europe. This last type of regional delineation is what we will mainly utilize in our class.

## What is a Region?

The human intellect seems driven to make sense out of the universe by lumping together phenomena into categories of similarity. Biology has its taxa of living objects, history its eras and periods, geology its epochs of earth history. Geographers too make sense of the world by compressing and synthesizing vast amounts of information into spatial categories of similar traits. The resulting areal units are referred to as regions.

Sometimes the unifying threads of a region are physical, such as climate and vegetation, resulting in regional designations such as the Sahara Desert or the tropical rainforest. Other times the threads are economic and cultural, as in the popular term Midwest for the central United States. All human beings need to compress large amounts of information into some stereotypes; often the geographic region is just that, a spatial stereotype for a section of earth that has some special signature or characteristic which sets it apart from other places on earth.

That is a good jumping off point for what we will consider a region in this class. A region has three things:

(1) has to have some area

(2) has to have some boundaries--although these boundaries are typically fuzzy, or imperfectly defined, and

(3) it has to have some homogeneous trait which sets it apart from surrounding areas.

What trait is homogeneous is defined by the user. You can define any place on earth as being in an infinite number of regions, depending on what trait you pick. The campus of Virginia Tech is simultaneously in a regions we could define as in the town of Blacksburg, as in SW Virginia, as in the state of Virginia, as in SE United States, as in a forested region, as in a humid subtropical climate zone, as in the Mid Atlantic, as in North America, as in coal mining belt, as in a Bible belt, as in the Western Hemisphere--what region do you think we are in?

## A Matter of Scale

There is a sense of scale to all systematic inquiry, whatever the discipline. In biology, for example, some scientists study the smaller units of cells, genes or molecules while others take a larger view, examining plants, animals or ecosystems. Similarly, some historians may focus on a specific individual at specific point in time, while others take a broader view of international events changing over several decades or throughout a century. Geographers also work at different scales. While some may concentrate on analysis of a local landscape--like the geography of Blacksburg, or the Blue Ridge Mountains--others will focus on the regional picture, on the spatial characteristics and interactions of a larger sphere, like the entire United States, or North America. Others do research at a still larger global scale, examining emerging trade networks between Asian countries and North America.

This ability to move between different scales--global, regional, local--is critical for understanding contemporary world regional geography because of the way globalization links together all people and places. Few villages today, however remote, are without contact to the modern world. Global economies draw upon local crops and resources, and conversely, fluctuations in world commodity prices affect the well-being of those local people. Global TV and video introduce foreign styles, ideas, mannerisms, and expectations into small towns and remote settlements in formerly isolated parts of the world. Further, very few places and peoples are isolated from global politics and tensions, from the influences of superpowers, supranational organizations, or, increasingly, from the ambitions and agendas of regional separatists and national splinter groups.

The other reason to pay attention to scale is because it plays an important component of our definition of regions. Since we have already expressed that regions have some sort of homogeneous factor which defines them, we must consider at what scale does this homogeneity apply--because it is defined by the scale itself. Let me give you an example.

Two presidential elections ago our country, by majority, elected Bill Clinton--a Democrat--to the presidency. Since more than half voted Democrat, we could say that the US at the country scale is a Democratic region, based on that singular homogeneous trait. However, if we looked at the state of Virginia, it voted predominately Republican--so at a smaller scale, we are in a Republican region. Montgomery County, a smaller region within Virginia, may have voted predominately Democrat. Maybe most of the people on your block, a smaller region still, voted Republican.

Perhaps everyone in your house voted Democrat, so you are back in a Democratic region. Thus, defining regions based on voting preferences DEMANDS that you state the scale of focus. And most importantly, the larger the region you define, the more exceptions to your homogeneous trait you will find within your region. This is what generalizations are all about--we are going to discuss and define our regions with GENERALLY HOMOGENEOUS traits within the region, knowing full well our generalization won't apply to everyone and every place.

## BRANCHES OF GEOGRAPHY

### Physical Geography

Physical geography (or physiogeography) focuses on geography as an Earth science. It aims to understand the physical features of the Earth, its lithosphere, hydrosphere, atmosphere, pedosphere and global flora and fauna patterns (biosphere). Physical Geography can be divided into the following broad categories:

## BIOGEOGRAPHY

Biogeography is the science which deals with geographic patterns of species distribution and the processes that result in such patterns.

The patterns of species distribution at this level can usually be explained through a combination of historical factors such as speciation, extinction, continental drift, glaciation (and associated variations in sea level, river routes, and so on), and river capture, in combination with the area and isolation of landmasses (geographic constraints) and available energy supplies.

### History

The theory of biogeography grows out of the work of Alfred Russel Wallace and other early evolutionary scientists. Wallace studied the distribution of flora and fauna in the Malay Archipelago in the 19th century. With the exception of Wallace and a few others, prior to the publication of The Theory of Island Biogeography by Robert MacArthur and E.O. Wilson in 1967 (which expanded their 1963 paper on the same topic) the field of biogeography was seen as a primarily historical one, and as such the field was seen as a purely descriptive one.

MacArthur and Wilson changed this perception, and showed that the species richness of an area could be predicted in terms of such factors as habitat area, immigration rate and extinction rate. This gave rise to an interest in island biogeography. The application of island biogeography theory to habitat fragments spurred the development of the fields of conservation biology and landscape ecology (at least among British and American academics; landscape ecology has a distinct genesis among European academics).

Classic biogeography has been expanded by the development of molecular systematics, creating a new discipline known as phylogeography. This development allowed scientists to test theories about the origin and dispersal of populations, such as island endemics. For example, while classic biogeographers were able to speculate about the origins of species in the Hawaiian Islands, phylogeography allows them to test theories of relatedness between these populations and putative source populations in Asia and North America.

Paleobiogeography goes one step further to include paleogeographic data and considerations of plate tectonics. Using molecular analyses and corroborated by fossils, it has been possible to demonstrate that perching birds evolved first in the region of Australia or the adjacent Antarctic (which at that time lay somewhat further north and had a temperate climate). From there, they spread to the other Gondwanan continents and Southeast Asia - the part of Laurasia then closest to their origin of dispersal - in the late Paleogene, before achieving a global distribution in the early Neogene (Jonsson & Fjeldsa 2006).

Not knowing the fact that at the time of dispersal, the Indian Ocean was much narrower than it is today, and that South America was closer to the Antarctic, one would be hard pressed to explain the presence of many "ancient" lineages of perching birds in Africa, as well as the mainly South American distribution of the suboscines.

## Classification

Biogeography is a synthetic science, related to geography, biology, soil science, geology, climatology, ecology and evolution.

Some fundamentals in biogeography are:

- evolution (change in genetic composition of a population)
- extinction (disappearance of a species)
- dispersal (movement of populations away from their point of origin, related to migration)

- range and distribution
- endemic areas
- vicariance

## CLIMATOLOGY

Climatology is the study of climate, scientifically defined as weather conditions averaged over a period of time, and is a branch of the atmospheric sciences. Basic knowledge of climate can be used within shorter term weather forecasting using analog techniques such as teleconnections and climate indices.

### Differences with Meteorology

In contrast to meteorology, which studies short term weather systems lasting up to a few weeks, climatology studies the frequency and trends of those systems. It studies the periodicity of weather events over years to millennia, as well as changes in long-term average weather patterns, in relation to atmospheric conditions. Climatologists, those who practice climatology, study both the nature of climates - local, regional or global - and the natural or human-induced factors that cause climates to change. Climatology considers the past and can help predict future climate change.

Phenomena of climatological interest include the atmospheric boundary layer, circulation patterns, heat transfer (radiative, convective and latent), interactions between the atmosphere and the oceans and land surface (particularly vegetation, land use and topography), and the chemical and physical composition of the atmosphere. Related disciplines include astrophysics, atmospheric physics, chemistry, ecology, geology, geophysics, glaciology, hydrology, oceanography, and volcanology.

### History

Perhaps the earliest person to hypothesize the concept of climate change was the medieval Chinese scientist Shen Kuo (1031-1095 AD). Shen Kuo theorized that climates naturally shifted over an enormous span of time, after observing petrified bamboos found underground near Yanzhou (modern day Yan'an,

Shaanxi province), a dry climate area unsuitable for the growth of bamboos. Early climate researchers include Edmund Halley, who published a map of the trade winds in 1686, after a voyage to the southern hemisphere. Benjamin Franklin, a renaissance man in the 18th century, was the first to map the course of the Gulf Stream for use in sending mail overseas from the United States to Europe. Francis Galton invented the term anticyclone. Helmut Landsberg led to statistical analysis being used in climatology, which led to its evolution into a physical science.

### Different Approaches

Climatology is approached in a variety of ways. Paleoclimatology seeks to reconstruct past climates by examining records such as ice cores and tree rings (dendroclimatology). Paleotempestology uses these same records to help determine hurricane frequency over millennia. The study of contemporary climates incorporates meteorological data accumulated over many years, such as records of rainfall, temperature and atmospheric composition. Knowledge of the atmosphere and its dynamics is also embodied in models, either statistical or mathematical, which help by integrating different observations and testing how they fit together. Modeling is used for understanding past, present and potential future climates. Historical climatology is the study of climate as related to human history and thus focuses only on the last few thousand years.

Climate research is made difficult by the large scale, long time periods, and complex processes which govern climate. Climate is governed by physical laws that can be expressed as differential equations. These equations are coupled and nonlinear, so that approximate solutions are obtained by using numerical methods to create global climate models. Climate is sometimes modeled as a stochastic process but this is generally accepted as an approximation to processes that are otherwise too complicated to analyze.

### Use in Weather Forecasting

A more complicated way of making a forecast, the analog technique requires remembering a previous weather event which is expected to be mimicked by an upcoming event. What makes it a difficult technique to use is that there is rarely a perfect analog for an event in the future. Some call this type of forecasting pattern recognition, which remains a useful method of observing rainfall over data voids such as oceans with knowledge of how satellite imagery relates to precipitation rates over land, as well as the forecasting of precipitation amounts and distribution in the future. A variation on this theme is used in Medium Range forecasting, which is known as teleconnections, when you use systems in other locations to help pin down the location of another system within the surrounding regime. One method of using teleconnections are by using climate indices such as ENSO-related phenomena.

### Climate Indices

Scientists use climate indices in their attempt to characterize and understand the various climate mechanisms that culminate in our daily weather. Much in the way the Dow Jones Industrial Average, which is based on the stock prices of 30 companies, is used to represent the fluctuations in the stock market as a whole, climate indices are used to represent the essential elements of climate. Climate indices are generally devised with the twin objectives of simplicity and completeness, and each index typically represents the status and timing of the climate factor it represents. By their very nature, indices are simple, and combine many details into a generalized, overall description of the atmosphere or ocean which can be used to characterize the factors which impact the global climate system.

### El Nino - Southern Oscillation

El Niño-Southern Oscillation (ENSO) is a global coupled ocean-atmosphere phenomenon. The Pacific ocean signatures, El Niño and La Niña are important temperature fluctuations in surface waters of the tropical Eastern Pacific Ocean. The name El Niño, from the Spanish for "the little boy", refers to the Christ child, because the phenomenon is usually noticed around Christmas time in the Pacific Ocean off the west coast of South America. La Niña means "the little girl". Their effect on climate in the subtropics and the tropics are profound. The atmospheric signature, the Southern Oscillation (SO) reflects the monthly or seasonal fluctuations in the air pressure difference between Tahiti and Darwin. The most recent occurrence of El Niño started in September 2006 and lasted until early 2007.

ENSO is a set of interacting parts of a single global system of coupled ocean-atmosphere climate fluctuations that come about as a consequence of oceanic and atmospheric circulation. ENSO is the most prominent known source of inter-annual variability in weather and climate around the world (~3 to 8 years), though not all areas are affected. ENSO has signatures in the Pacific, Atlantic and Indian Oceans. El Niño causes weather patterns which causes it to rain in specific places but not in others, this is one of many causes for the drought.

In the Pacific, during major warm events, El Niño warming extends over much of the tropical Pacific and becomes clearly linked to the SO intensity. While ENSO events are basically in phase between the Pacific and Indian Oceans, ENSO events in the Atlantic Ocean lag behind those in the Pacific by 12 to 18 months.

Many of the countries most affected by ENSO events are developing countries within main continents (South America, Africa...), with economies that are largely dependent upon their agricultural and fishery sectors as a major source of food supply, employment, and foreign exchange. New capabilities to predict the onset of ENSO events in the three oceans can have global socio-economic impacts. While ENSO is a global and natural part of the Earth's climate, whether its intensity or frequency may change as a result of global warming is an important concern. Low-frequency variability has been evidenced: the quasi-decadal oscillation (QDO). Inter-decadal (ID) modulation of ENSO (from PDO or IPO) might exist. This could explain the so-called protracted ENSO of the early 90s.

## Madden-Julian Oscillation

The Madden-Julian Oscillation (MJO) is an equatorial traveling pattern of anomalous rainfall that is planetary in scale. It is characterized by an eastward progression of large regions of both enhanced and suppressed tropical rainfall, observed mainly over the Indian Ocean and Pacific Ocean. The anomalous rainfall is usually first evident over the western Indian Ocean, and remains evident as it propagates over the very warm ocean waters of the western and central tropical Pacific. This pattern of tropical rainfall then generally becomes very nondescript as it moves over the cooler ocean waters of the eastern Pacific but reappears over the tropical Atlantic and Indian Ocean. The wet phase of enhanced convection and precipitation is followed by a dry phase where convection is suppressed. Each cycle lasts approximately 30-60 days. The MJO is also known as the 30-60 day oscillation, 30-60 day wave, or intraseasonal oscillation.

### North Atlantic Oscillation (NAO)

Indices of the NAO are based on the difference of normalized sea level pressure (SLP) between Ponta Delgada, Azores and Stykkisholmur/Reykjavik, Iceland. The SLP anomalies at each station were normalized by division of each seasonal mean pressure by the long-term mean (1865-1984) standard deviation. Normalization is done to avoid the series of being dominated by the greater variability of the northern of the two stations. Positive values of the index indicate stronger-than-average westerlies over the middle latitudes.

### Northern Annualar Mode (NAM) or Arctic Oscillation (AO)

The NAM, or AO, is defined as the first EOF of northern hemisphere winter SLP data from the tropics and subtropics. It explains 23% of the average winter (December-March) variance, and it is dominated by the NAO structure in the Atlantic. Although there are some subtle differences from the regional pattern over the Atlantic and Arctic, the main difference is larger amplitude anomalies over the North Pacific of the same sign as those over the Atlantic. This feature gives the NAM a more annular (or zonally-symmetric) structure.

### Northern Pacific (NP) Index

The NP Index is the area-weighted sea level pressure over the region 30°N-65°N, 160°E-140°W.

## Pacific Decadal Oscillation (PDO)

The PDO is a pattern of Pacific climate variability that shifts phases on at least inter-decadal time scale, usually about 20 to 30 years. The PDO is detected as warm or cool surface waters in the Pacific Ocean, north of 20°N. During a "warm", or "positive", phase, the west Pacific becomes cool and part of the eastern ocean warms; during a "cool" or "negative" phase, the opposite pattern occurs. The mechanism by which the pattern lasts over several years has not been identified; one suggestion is that a thin layer of warm water during summer may shield deeper cold waters. A PDO signal has been reconstructed to 1661 through tree-ring chronologies in the Baja California area.

## Interdecadal Pacific Oscillation (IPO)

The Interdecadal Pacific Oscillation (IPO or ID) display similar sea surface temperature (SST) and sea level pressure patterns to the PDO, with a cycle of 15-30 years, but affects both the north and south Pacific. In the tropical Pacific, maximum SST anomalies are found away from the equator. This is quite different to the quasi-decadal oscillation (QDO) with a period of 8-to-12 years and maximum SST anomalies straddling the equator, thus resembling ENSO.

## Climate Models

Climate models use quantitative methods to simulate the interactions of the atmosphere, oceans, land surface, and ice. They are used for a variety of purposes from study of the dynamics of the weather and climate system to projections of future climate. All climate models balance, or very nearly balance, incoming energy as short wave (including visible) electromagnetic radiation to the earth with outgoing energy as long wave (infrared) electromagnetic radiation from the earth. Any unbalance results in a change in the average temperature of the earth.

The most talked-about models of recent years have been those relating temperature to emissions of carbon dioxide (see greenhouse gas). These models predict an upward trend in the surface temperature record, as well as a more rapid increase in temperature at higher altitudes. Models can range from relatively simple to quite complex:

- A simple radiant heat transfer model that treats the earth as a single point and averages outgoing energy
- this can be expanded vertically (radiative-convective models), or horizontally
- finally, (coupled) atmosphere-ocean-sea ice global climate models discretise and solve the full equations for mass and energy transfer and radiant exchange.

## PALEOCLIMATOLOGY

Paleoclimatology is the study of climate change taken on the scale of the entire history of Earth. It uses records from ice sheets, tree rings, sediment, and rocks to determine the past state of the climate system on Earth.

### Techniques of Paleoclimatology

Paleoclimatologists employ a wide variety of skills to arrive at their theories and conclusions. Glaciers and the polar ice caps/ice sheets are a widely employed source of data in paleoclimatology. Recent ice coring projects in the ice caps of Greenland and Antarctica have yielded data going back several hundred thousand years -- over 800,000 years in the case of the EPICA project.

- Inside of these layers scientists have found pollen, allowing them to estimate the total amount of plant growth of that year by the pollen count. The thickness of the layer can help to determine the amount of precipitation that year. Certain layers contain ash from volcanic eruptions.
- Air trapped within fallen snow becomes encased in tiny bubbles as the snow is compressed into ice in the glacier under the weight of later years' snow. This trapped air has proven a tremendously valuable source for direct measurement of the composition of air from the time the ice was formed.
- Because evaporation rates of water molecules with slightly heavier isotopes of hydrogen and oxygen are

slightly different during warmer and colder periods, changes in the average temperature of the ocean surface are reflected in slightly different ratios between those isotopes. Various cycles in those isotope ratios have been detected.

Dendroclimatology is the science of extracting climate information from tree rings. Rings from living trees of great age give data about recent centuries back to a few millenia. Older intact wood that has escaped decay can extend the time covered by identifying patterns that match rings of known age from live trees. Petrified tree rings give paleoclimatology data over a much larger stretch of time. The fossil itself is dated with radioactive dating within a wide margin of error. The rings themselves can give some information about rainfall and temperature during that epoch.

Sediment layers have been studied, particularly those in the bottom of lakes and oceans. Characteristics of preserved vegetation, animals, pollen, and isotope ratios provide information. The study of pollens, known as palynology provides evidence of the abundance of different plant species, a proxy indicator of climate.

Sedimentary rock layers provide a more compressed view of climate, as the layers of sedimentary rocks span hundreds of thousands to millions of years. Scientists can get a grasp of long term climate by studying sedimentary rock going back billions of years. The division of earth history into separate periods is largely based on visible changes in sedimentary rock layers that demarcate major changes in conditions. Often these include major shifts in climate.

**Planet's Timeline**

Some of the mile stones that mark the history of the planet are as follows (Ma = Millions of years ago):

| | |
|---|---|
| 4,000 Ma | earliest biogenic carbon |
| 3,800 Ma | banded iron formations (with reduced iron) |
| 3,700 Ma | oldest rocks |
| 3,500 Ma | oldest stromatolites |

| | |
|---|---|
| 3,450 Ma | earliest bacteria |
| 3,000 Ma | earliest precambrian ice ages |
| [?] | Chuos Tillites of South-West Africa |
| 3,000 Ma | earliest photosynthetic bacteria |
| 2,700 Ma | oldest chemical evidence of complex cells |
| 2,300 Ma | first green algae (eukaryotes) |
| 2,000 Ma | free oxygen in the atmosphere |
| 2,000 Ma to 1600 Ma | Gowganda tillites in the Canadian shield |
| 1,700 Ma | end of the banded iron formations and red beds become abundant (non-reducing atmosphere) |
| 750 Ma | Sturtian tillites in southern Australia indicate major glaciation |
| 700 Ma | first metazoans late Proterozoic (Ediacaran Epoch) - first skeletons |
| 570 Ma to present | Phanerozoic Eon |
| 100 Ma | development of the angiosperms (flowering plants) |
| 2 Ma to present | modern world and man's appearance on earth |
| 0.01 Ma | end of the last ice age |
| 0.001 Ma | warming trend of the middle ages |
| 0.0001 Ma | end of the mini ice age |
| 0.00022 Ma to present | industrialized human world. |

## HISTORY OF THE ATMOSPHERE

### Earliest Atmosphere

The earliest atmosphere of the Earth was probably stripped away by solar winds early in the history of the planet. These gases were later replaced by an atmosphere derived from outgassing from the Earth. Sometime during the late Archean Era an oxygen atmosphere began to develop from photosynthesizing algae.

### *Carbon Dioxide and Free Oxygen*

Free oxygen did not exist until about 1,700 Ma and this can be seen with the development of the red beds and the end of the banded iron formations. This signifies a shift from a reducing atmosphere to an oxidising atmosphere. The early atmosphere and hydrosphere (up until about 2,000 Ma) were devoid of free Oxygen. After photosynthesis developed, photoautotrophs began releasing $O_2$.

The very early atmosphere of the earth contained mostly carbon dioxide ($CO_2$): about 80%. This gradually dropped to about 20% by 3,500 Ma. This coincides with the development of the first bacteria about 3,500 Ma. By the time of the development of photosynthesis (2,700 Ma), $CO_2$ levels in the atmosphere were in the range of 15%. During the period from about 2,700 Ma to about 2,000 Ma, photosynthesis dropped the $CO_2$ concentrations from about 15% to about 8%. By about 2,000 Ma free $O_2$ was beginning to accumulate. This gradual reduction in $CO_2$ levels continued to about 600 Ma at which point $CO_2$ levels were below 1% and $O_2$ levels had risen to more than 15%. 600Ma corresponds to the end of the Precambrian and the beginning of the Cambrian, the end of the cryptozoic and the beginning of the Phanerozic, and the beginning of oxygen-breathing life.

### *Precambrian Climate*

The climate of the late Precambrian was typically cold with glaciation spreading over much of the earth. At this time the continents were bunched up in a supercontinent called Rodinia. Massive deposits of tillites are found and anomalous isotopic signatures are found which are consistent with the idea that the earth at this time was a massive snowball. Map of Rodinia at the end of the Precambrian after Australia and Antarctica rotated away from the southern hemisphere. As the Proterozoic Eon drew to a close, the Earth started to warm up. By the dawn of the Cambrian and the Phanerozoic Eon, Earth was experiencing average global temperatures of about +22 °C. Hundreds of millions of years of ice were replaced with the balmy tropical seas of the Cambrian Period within which life exploded at a rate never seen before or after.

### *Phanerozoic Climate*

Qualitatively, the Earth's climate was varied between conditions that support large-scale continental glaciation and those which are extensively tropical and lack permanent ice caps even at the poles. The time scale for this variation is roughly 140 million years and may be related to Earth's motion into and out of galactic spiral arms (Veizer and Shaviv 2003).

The difference in global mean temperatures between a fully glacial earth and ice free Earth is estimated at approximately 10 °C, though far larger changes would be observed at high latitudes and smaller ones at low latitudes. One key requirement for the development of large scale ice sheets is the arrangement of continental land masses at or near the poles. With plate tectonics constantly rearranging the continents, it can also shape long-term climate evolution. However, the presence of land masses at the poles is not sufficient to guarantee glaciations. Evidence exists of past warm periods in Earth's climate when polar land masses similar to Antarctica were home to deciduous forests rather than ice sheets.

Changes in the atmosphere may also exert an important influence over climate change. The establishment of $CO_2$-consuming (and oxygen-producing) photosythesizing organisms in the Precambrian led to the production of an atmosphere much like today's, though for most of this period it was much higher in $CO_2$ than today. Similarly, the Earth's average temperature was also frequently higher than at present, though it has been argued that over very long time scales climate is largely decoupled from carbon dioxide variations (Veizer et al. 2000). Or more specifically that changing continental configurations and mountain building probably have a larger impact on climate than carbon dioxide. Others dispute this, and suggest that the variations of temperature in response to carbon dioxide changes have been underestimated (Royer et al. 2004). However, it is clear that the preindustrial atmosphere with only 280 ppm $CO_2$ is not far from the lowest ever occurring since the rise of macroscopic life.

Superimposed on the long-term evolution between hot and cold climates have been many short-term fluctuations in climate similar to, and sometimes more severe than, the varying glacial and interglacial states of the present ice age. Some of the most severe fluctuations, such as the Paleocene-Eocene Thermal

Maximum, may be related to rapid increases in atmospheric carbon dioxide due to the collapse of natural methane reservoirs in the oceans (see methane clathrates). Severe climate changes also seem to have occurred during the course of the Cretaceous-Tertiary, Permian-Triassic, and Ordovician-Silurian extinction events; however, it is unclear to what degree these changes caused the extinctions rather than merely responding to other processes that may have been more directly responsible for the extinctions.

### *Quaternary Sub-era*

The Quaternary sub-era includes the current climate. There has been a cycle of ice ages for the past 2.2-2.1 million years (starting before the Quaternary in the late Neogene Period).

Ice core data for the past 400,000 years. Note length of glacial cycles averages ~100,000 years. Blue curve is temperature, green curve is $CO_2$, and red curve is windblown glacial dust (loess).

### Controlling Factors

Geologically short-term (<120,000 year) temperatures are believed to be driven by orbital factors (see Milankovitch cycles). The arrangements of land masses on the Earth's surface are believed to reinforce these orbital forcing effects.

Continental drift obviously affects the thermohaline circulation, which transfers heat between the equatorial regions and the poles, as does the extent of polar ice coverage.

The timing of ice ages throughout geologic history is in part controlled by the position of the continental plates on the surface of the Earth. When landmasses are concentrated near the polar regions, there is an increased chance for snow and ice to accumulate. Small changes in solar energy can tip the balance between summers in which the winter snow mass completely melts and summers in which the winter snow persists until the following winter.

Comparisons of plate tectonic continent reconstructions and paleoclimatic studies show that the Milankovitch cycles have the greatest effect during geologic eras when landmasses have been concentrated in polar regions, as is the case today. Today, Greenland, Antarctica, and the northern portions of Europe, Asia, and North America are situated such that a minor change in solar energy will tip the balance between year-round snow/ice preservation and complete summer melting. The presence of snow and ice is a well-understood positive feedback mechanism for climate. The Earth today is considered to be prone to ice age glaciations.

Another proposed factor in long term temperature change is the Uplift-Weathering Hypothesis, first put forward by T. C. Chamberlin in 1899 and later independently proposed in 1988 by Maureen Raymo and colleagues, where upthrusting mountain ranges expose minerals to weathering resulting in their chemical conversion to carbonates thereby removing CO2 from the atmosphere and cooling the earth. Others have proposed similar effects due to changes in average water table levels and consequent changes in sub-surface biological activity and PH levels.

Over the very long term the energy output of the sun has gradually increased, on the order of 5% per billion (109) years, and will continue to do so until it reaches the end of its current phase of stellar evolution.

2

# Coastal Geography

Coastal geography is the study of the dynamic interface between the ocean and the land, incorporating both the physical geography (i.e coastal geomorphology, geology and oceanography) and the human geography (sociology and history) of the coast. It involves an understanding of coastal weathering processes, particularly wave action, sediment movement and weather, and also the ways in which humans interact with the coast.

## WAVE ACTION AND LONGSHORE DRIFT

The waves of different strengths that constantly hit against the shoreline are the primary movers and shapers of the coastline. Despite the simplicity of this process, the differences between waves and the rocks they hit result in hugely varying shapes.

The effect that waves have depends on their strength. Strong, also called destructive waves occur on high energy beaches and are typical of Winter. They reduce the quantity of sediment present on the beach by carrying it out to bars under the sea. Constructive, weak waves are typical of low energy beaches and occur most during summer. They do the opposite to destructive waves and increase the size of the beach by piling sediment up onto the berm.

One of the most important transport mechanisms results from wave refraction. Since waves rarely break onto a shore at right angles, the upward movement of water onto the beach (swash) occurs at an oblique angle. However, the return of water (backwash) is at right angles to the beach, resulting in the net movement of beach material laterally. This movement is known as beach drift. The endless cycle of swash and backwash and resulting beach drift can be observed on all beaches.

The wide expanse of sand and low waves, even though a storm is blowing onshore, indicate that this coastline at Rhossili in Wales is a low energy shoreline.

Probably the most important effect is longshore drift (LSD), the process by which sediment is continuously moved along beaches by wave action. LSD occurs because waves hit the shore at an angle, pick up sediment (sand) on the shore and carry it down the beach at an angle (this is called swash). Due to gravity, the water then falls back perpendicular to the beach, dropping its sediment as it loses energy (this is called backwash). The sediment is then picked up by the next wave and pushed slightly further down the beach, resulting in a continual movement of sediment in one direction. This is the reason why long strips of coast are covered in sediment, not just the areas around river mouths, which are the main sources of beach sediment. LSD is reliant on a constant supply of sediment from rivers and if sediment supply is stopped or sediment falls into a submarine canals at any point along a beach, this can lead to bare beaches further along the shore.

LSD helps create many landforms including barriers, bay beaches and spits. In general LSD action serves to straighten the coast because the creation of barriers cuts off bays from the sea while sediment usually builds up in bays because the waves there are weaker (due to wave refraction), while sediment is carried away from the exposed headlands. The lack of sediment on headlands removes the protection from waves them and makes them more vulnerable to weathering while the gathering of sediment in bays (where longshore drift is unable to remove it) protects the bays from further erosion and makes them pleasant recreational beaches.

**Atmospheric Processes**

- Onshore winds blowing "up" the beach, pick up sand and moves it up the beach to form sand dunes.
- Rain hits the shore and erodes rocks and carries weathered material to the shoreline to form beaches.

- Warm weather can encourage chemical and biological processes to occur. In tropical areas some plants and animals protect stones from weathering, while others actually eat away at the rocks.
- Temperatures that vary from below to above freezing point result in freeze-thaw weathering, while weather more than a few degrees below freezing point creates sea ice.

### Biological Processes

In tropical regions in particular, plants and animals not only affect the weathering of rocks but are a source of sediment themselves. The shells and skeletons of many organisms are of calcium carbonate and when this is broken down it forms sediment, limestone and clay.

### Physical Processes

The main Physical Weathering process on beaches is salt-crystal growth. Wind carries salt spray onto rocks, where it is absorbed into small pores and cracks within the rocks. There the water evaporates and the salt crystallises, creating pressure and often breaking down the rock. In some beaches calcium carbonate is able to bind together other sediments to form beachrock and in warmer areas dunerock.

## SEA LEVEL CHANGES

Because the sea level on earth regularly rises and falls due to climatic changes. During cold periods more of the Earth's water is stored as ice in glaciers while during warm periods it is released and sea levels rise to cover more land. Sea levels are currently quite high, while just 18,000 years ago during the Pleistocene ice age they were quite low. Global warming may result in further rises in the future, which presents a risk to coastal cities as most would be flooded by only small rises. As sea levels rise fjords and rias form. Fjords are flooded glacial valleys and rias are flooded river valleys. Fjords typically have steep rocky sides, while rias have dendritic drainage patterns typical of drainage zones. As tectonic plates move about the Earth they can rise and fall due to changing pressures and the presence of glaciers. If a beach is moving upwards relative to other plates this is known as isostatic change and raised beaches can be formed.

## COASTAL LANDFORMS

### Spits

If the coast suddenly changes direction, especially around an estuary, spits are likely to form. LSD pushes sediment along the beach but when it reaches a turn as in the diagram, the LSD does not always easily turn with it, especially near an estuary where the outward flow from a river may push sediment away from the coast. The area may be also be shielded from wave action, preventing much LSD. On the side of the headland receiving weaker waves, shingle and other large sediments will build up under the water where waves are not strong enough to move them along. This provides a good place for smaller sediments to build up to sea level. The sediment, after passing the headland will accumulate on the other side and not continue down the beach, sheltered both by the headland and the shingle.

Slowly over time sediment simply builds on this area, extending the spit outwards, forming a barrier of sand. Once in a while, the wind direction will change and come from the other direction. During this period the sediment will be pushed along in the other direction. The spit will start to grow backwards, forming a 'hook'. After this time the spit will grow again in the original direction. Eventually the spit will not be able to grow any further because it is no longer sufficiently sheltered from erosion by waves, or because the estuary current prevents sediment resting. Usually in the salty but calm waters behind the spit there will form a salt marshland. Spits often form around the breakwater of artificial harbours requiring dredging.

Occasionally, if there is no estuary then it is possible for the spit to grow across to the other side of the bay and form what is called a bar, or barrier. Barriers come in several varieties, but all form in a manner similar to spits. They usually enclose a bay to form a lagoon. They can join two headlands or join a headland to the mainland. When an island is joined to the mainland with a bar or barrier it is known as a tombolo. This usually occurs due to wave refraction, but can also be caused by isostatic change, a change in the level of the land (e.g. Chesil Beach). An example of this is along the Holderness coastline.

## COASTAL EROSION

Many stretches of the coastline of East Anglia, England, are prone to high rates of erosion, as illustrated by this collapsed section of the cliffs at Hunstanton, Norfolk.

Coastal erosion see also (beach evolution) is the wearing away of land or the removal of beach or dune sediments by wave action, tidal currents, wave currents, or drainage. Waves, generated by storms or fast moving moter craft, cause coastal erosion, which may take the form of long-term losses of sediment and rocks, or merely in the temporary redistribution of coastal sediments; erosion in one location may result in accretion nearby. The study of erosion and sediment redistribution is called 'coastal morphodynamics'.

On rocky coasts, coastal erosion results in dramatic rock formations in areas where the coastline contains rock layers or fracture zones with different resistances to erosion. The softer areas become eroded much faster than the harder ones, which can result in typical landforms such as tunnels, bridges, columns, and pillars.

## EFFECT ON HUMAN ESTABLISHMENTS

On sedimentary coasts, coastal erosion typically poses more of a danger to human settlements than it does to nature itself. Dunwich, the capital of the English medieval wool trade, disappeared over the space of a few centuries due to redistribution of sediment by waves. Human interference can also increase coastal erosion: Hallsands in Devon, England, was a coastal village that was washed away overnight, an event possibly exacerbated by dredging of shingle in the bay in front of it.

The California coast, which has soft cliffs of sedimentary rock and is heavily populated, regularly has incidents of housing damage as cliffs erode. Damage in Pacifica is shown at right. Devil's Slide, Santa Barbara and Malibu are regularly affected.

The Holderness coastline on the east coast of England, just north of the Humber Estuary, is the fastest eroding coastline in Europe due to its soft clay cliffs and powerful waves. Groynes and other artificial measure to keep it under control has only sped up the process further down the coast, because longshore drift starves the beaches of sand, leaving them more exposed.

## Wave Action

The ability of waves to cause erosion of the cliff face depends on number of factors, which include:

- The hardness or 'erodibility' of the rocks exposed at the base of the cliff
  * The key factors in determining erodibility include the rock strength along with the presence of fissures, fractures, and beds of non-cohesive materials such as silt and fine sand.
- The rate at which cliff fall debris is removed from the foreshore
  * Debris removal from the foreshore is dependent on the power of the waves crossing the beach, and this energy must reach a critical level or material will not be removed from the debris lobe. On many cliffs these debris lobes can be very persistent and may take many years before they are completely removed.
- The presence or absence of a beach at the base of the cliff
  * Beaches help dissipate wave energy on the foreshore and can provide a measure of protection to the cliff from marine erosion.
- The stability of the foreshore, or its resistance to lowering
  * Lowering of the beach or shore platform through wave action is a key factor controlling the rate of cliff recession. If the beach is not lowered the foreshore should widen and become more effective at dissipating the wave energy, so that fewer and less powerful waves reach the cliff.
- The adjacent bathymetry

* The nearshore bathymetry controls the wave energy arriving at the coast, and can have an important influence on the rate of cliff erosion.
* The supply of beach material in the coastal cell from updrift
* The provision of material eroded updrift coming onto the foreshore beneath the cliff will help ensure a stable beach, thus providing a measure of protection.

## RATE OF EROSION

The factors which control the rate of erosion:

| First Order (Most Important) | Second Order | Third Order |
|---|---|---|
| Geological structure | Weathering and | |
| and lithology | transport slope processes | |
| - Hardness | Slope hydrology | |
| - Height etc | Vegetation | |
| - Fractures / faults | Cliff foot erosion | |
| Wave climate | Cliff foot sediment accumulation | |
| - Prevailing wave direction | Resistance of cliff foot sediment | |
| Sub-aerial climate | to attrition and transport | |
| - Weathering (frost etc | | |
| - Stress relief swelling / shrinkage | | |
| Water-level change | | |
| - Groundwater fluctuations | | |
| - Tidal range | | |
| Geomorphology | | |

## WEATHERING

Weathering is the process of breaking down rocks, soils and their minerals through direct contact with the atmosphere. Weathering occurs in situ, or 'without movement', and thus should not to be confused with erosion, which involves the movement and disintegration of rocks and minerals by processes such as water, wind, ice and gravity. Two main classifications of weathering processes exist. Mechanical or physical weathering involves the breakdown of rocks and soils through direct contact with atmospheric conditions such as heat, water, ice and pressure. The second classification, chemical weathering, involves the direct effect of atmospheric chemicals, or biologically produced chemicals (also known as biological weathering), in the breakdown of rocks, soils and minerals.

The materials left over after the rock breaks down combined with organic material creates soil. The mineral content of the soil is determined by the parent material, thus a soil derived from a single rock type can often be deficient in one or more minerals for good fertility, while a soil weathered from a mix of rock types (as in glacial, eolian or alluvial sediments) often makes more fertile soil.

## Physical (mechanical) Weathering

Mechanical weathering is a cause of the disintegration of rocks. The primary process in mechanical weathering is abrasion - the process by which clasts and other particles are reduced in size. However, chemical and physical weathering often go hand in hand. For example, cracks exploited by mechanical weathering will increase the surface area exposed to chemical action. Furthermore, the chemical action at minerals in cracks can aid the disintegration process.

## Thermal Expansion

Thermal expansion, also known as onion-skin weathering, exfoliation, insulation weathering or thermal shock, often occurs in areas, like deserts, where there is a large diurnal temperature range. The temperatures soar high in the day, while dipping greatly at night. As the rock heats up and expands by day, and cools and contracts by night, stress is often exerted on the outer layers. The stress causes the peeling off of the outer layers of rocks in thin sheets. Though this is caused mainly by temperature changes, thermal expansion is enhanced by the presence of moisture.

## Freeze thaw Weathering

This process can also be called frost shattering. This type of weathering is common in mountain areas where the temperature is around freezing point. Frost induced weathering, although often attributed to the expansion of freezing water captured in cracks, is generally independent of the water-to-ice expansion. It has long been known that moist soils expand or frost heave upon freezing as a result of water migrating along from unfrozen areas via thin films to collect at growing ice lenses. This same phenomena occurs within pore spaces of rocks. They grow larger as they attract liquid water from the surrounding pores.

The ice crystal growth weakens the rocks which, in time, break up. Intermolecular forces acting between the mineral surfaces, ice, and water sustain these unfrozen films which transport moisture and generate pressure between mineral surfaces as the lens aggregates. Experiments show that chalk, sandstone and limestone do not fracture at the nominal freezing temperature of water of slightly below 0°C, even when cycled or held at low temperature for extended periods, as one would expect if weathering resulted from the expansion of water as froze. For the more porous types of rocks, the temperature range critical for rapid, ice-lens-induced fracture is -3 to -6°C, significantly below freezing temperatures.

Freeze induced weathering action occurs mainly in environments where there is a lot of moisture, and temperatures frequently fluctuate above and below freezing point-that is, mainly alpine and periglacial areas. An example of rocks susceptible to frost action is chalk, which has many pore spaces for the growth of ice crystals. This process can be seen in Dartmoor where it results in the formation of tors.

### Frost Wedging

Formerly believed to be the dominant mode, ice wedging may still be a factor for weathering of nonporous rock, although recent research has demonstrated it less important than previously thought. Frost action, sometimes known as ice crystal growth, ice wedging, frost wedging or freeze-thaw occurs when water in cracks and joints of rocks freeze and expand. Water can exert pressures up to 21 megapascals (MPa) (2100 kgf/cm$^2$) at -22 °C. This pressure is often higher than the resistance of most rocks and causes the rock to shatter.

When water that has entered the joints freezes, the ice formed strains the walls of the joints and causes the joints to deepen and widen. This is because the volume of water expands by 9% when it freezes.

When the ice thaws, water can flow further into the rock. When the temperature drops below freezing point and the water freezes again, the ice enlarges the joints further.

Repeated freeze-thaw action weakens the rocks which, over time, break up along the joints into angular pieces. The angular rock fragments gather at the foot of the slope to form a talus slope (or scree slope). The splitting of rocks along the joints into blocks is called block disintegration. The blocks of rocks that are detached are of various shapes depending on rock structure.

### Pressure Release

In pressure release, also known as unloading, overlying materials (not necessarily rocks) are removed (by erosion, or other processes), which causes underlying rocks to expand and fracture parallel to the surface. Often the overlying material is heavy, and the underlying rocks experience high pressure under them, for example, a moving glacier. Pressure release may also cause exfoliation to occur.

Intrusive igneous rocks (e.g. granite) are formed deep beneath the earth's surface. They are under tremendous pressure because of the overlying rock material. When erosion removes the overlying rock material, these intrusive rocks are exposed and the pressure on them is released. The outer parts of the rocks then tend to expand. The expansion sets up stresses which cause fractures parallel to the rock surface to form. Over time, sheets of rock break away from the exposed rocks along the fractures. Pressure release is also known as "exfoliation" or "sheeting"; these processes result in batholiths and granite domes, an example of which is Dartmoor.

### Hydraulic Action

This is when water (generally from powerful waves) rushes into cracks in the rockface rapidly. This traps a layer of air at the bottom of the crack, compressing it and weakening the rock. When the wave retreats, the trapped air is suddenly released with explosive force. The explosive release of highly pressurised air cracks away fragments at the rockface and widens the crack itself.

### Salt-crystal Growth (Haloclasty)

Salt crystallisation or otherwise known as Haloclasty causes disintegration of rocks when saline (see salinity) solutions seep into cracks and joints in the rocks and evaporate, leaving salt crystals behind. These salt crystals expand as they are heated up, exerting pressure on the confining rock. Salt crystallisation may also take place when solutions decompose rocks (for example, limestone and chalk) to form salt solutions of sodium sulfate or sodium carbonate, of which the moisture evaporates to form their respective salt crystals. The salts which have proved most effective in disintegrating rocks are sodium sulfate, magnesium sulfate, and calcium chloride. Some of these salts can expand up to three times or even more. It is normally associated with arid climates where strong heating causes strong evaporation and therefore salt crystallisation. It is also common along coasts. An example of salt weathering can be seen in the honeycombed stones in sea walls.

### Biotic Weathering

Living organisms may contribute to mechanical weathering (as well as chemical weathering, see 'biological' weathering below). Lichens and mosses grow on essentially bare rock surfaces and create a more humid chemical microenvironment. The attachment of these organisms to the rock surface enhances physical as well as chemical breakdown of the surface microlayer of the rock. On a larger scale seedlings sprouting in a crevice and plant roots exert physical pressure as well as providing a pathway for water and chemical inlfitration. Burrowing animals and insects disturb the soil layer adjacent to the bedrock surface thus further increasing water and acid infiltration and exposure to oxidation processes.

### Chemical Weathering

Chemical weathering involves the change in the composition of rock, often leading to a 'break down' in its form.

**Dissolution**

Rainfall is naturally slightly acidic because atmospheric carbon dioxide dissolves in the rainwater producing weak carbonic acid. In unpolluted environments, the rainfall pH is around 5.6.

Acid rain occurs when gases such as sulphur dioxide and nitrogen oxides are present in the atmosphere. These oxides react in the rain water to produce stronger acids and can lower the pH to 4.5 or even 3.0. Sulfur dioxide, SO2, comes from volcanic eruptions or from fossil fuels, can become sulfuric acid within rainwater, which can cause solution weathering to the rocks on which it falls.

One of the most well-known solution weathering processes is carbonation, the process in which atmospheric carbon dioxide leads to solution weathering.

Carbonation occurs on rocks which contain calcium carbonate such as limestone and chalk. This takes place when rain combines with carbon dioxide or an organic acid to form a weak carbonic acid which reacts with calcium carbonate (the limestone) and forms calcium bicarbonate.

This process speeds up with a decrease in temperature and therefore is a large feature of glacial weathering.

The reactions as follows:

$CO_2 + H_2O \rightarrow H_2CO_3$

carbon dioxide + water → carbonic acid

$H_2CO_3 + CaCO_3 \rightarrow Ca(HCO_3)_2$

carbonic acid + calcium carbonate → calcium bicarbonate

**Hydration**

Hydration is a form of chemical weathering that involves the rigid attachment of $H^+$ and $OH^-$ ions to the atoms and molecules of a mineral. When rock minerals take up water, the increased volume creates physical stresses within the rock. For example iron oxides are converted to iron hydroxides and the hydration of anhydrite forms gypsum.

## Hydrolysis

Hydrolysis is a chemical weathering process affecting Silicate minerals. In such reactions, pure water ionizes slightly and reacts with silicate minerals. An example reaction:

$$Mg_2SiO_4 + 4H^+ + 4OH^- \rightarrow 2Mg^{2+} + 4OH^- + H4SiO_4$$

olivine + four ionized → ions in + silicic acid
(forsterite) water molecules solution in solution

This reaction results in complete dissolution of the original mineral, assuming enough water is available to drive the reaction. However, the above reaction is to a degree deceptive because pure water rarely acts as a $H^+$ donor. Carbon dioxide, though, dissolves readily in water forming a weak acid and H+ donor.

$$Mg_2SiO_4 + 4CO_2 + 4H_2O \rightarrow 2Mg^{2+} + 4HCO_3^- + 4H_4SiO_4$$

olivine (forsterite) + carbon dioxide + water → Magnesium and bicarbonate ions in solution + silicic acid in solution

This hydrolysis reaction is much more common. Carbonic acid is consumed by silicate weathering, resulting in more alkaline solutions because of the bicarbonate. This is an important reaction in controlling the amount of $CO_2$ in the atmosphere and can affect climate.

Aluminosilicates when subjected to the hydrolysis reaction produce a secondary mineral rather than simply releasing cations.

$$2KAlSi_3O_8 + 2H_2CO_3 + 9H_2O \rightarrow Al_2Si_2O_5(OH)_4 + 4H_4SiO_4 + 2K^+ + 2HCO_3^-$$

Orthoclase (aluminosilicate feldspar) + carbonic acid + water → Kaolinite (a clay mineral) + silicic acid in solution + potassium and bicarbonate ions in solution

## Oxidation

Within the weathering environment chemical oxidation of a variety of metals occurs. The most commonly observed is the oxidation of $Fe^{2+}$ (iron) and combination with oxygen and water to form $Fe^{3+}$ hydroxides and oxides such as goethite, limonite, and hematite. This gives the affected rocks a reddish-brown coloration on the surface which crumbles easily and weakens the rock. This process is better known as 'rusting'.

important to note that it is the chemical combination of minerals with oxygen and water to form oxide or hydroxide, common in iron compounds and most effective in warm climates on igneous rocks that contain iron or manganese. Rusting of cars/iron railings. Fe+++ + e- Fe ++ iron electron iron oxide (rust).

### *Biological*

A number of plants and animals may create chemical weathering through release of acidic compounds.

The most common form of biological weathering is the release of chelating compounds, that is acids, by trees so as to break down elements such as aluminium and iron in the soils beneath them. Once broken down, such elements are more easily washed away by rainwater. This process exists as metals such as iron can be toxic and hinder the/a tree's growth. Extreme release of chelating compounds can easily affect surrounding rocks and soils, and may lead to podsolisation of soils.

## Building Weathering

Buildings made of limestone are particularly susceptible to weathering. Weeds grow almost anywhere without many problems. They can sometimes germinate in the gutters of buildings where they have been transported to by the wind. As they proceed to grow they plant their roots down into the rock that the building is made up of forcing their way further down. This causes the rock to exfoliate over a long time, small fragments crumbling away now and then. Statues and ornamental features can be badly damaged by weathering, especially in areas severely affected by acid rain which is caused by pollutants put into the air.

## Beach Evolution

The shoreline is where the land meets the sea and it is continually changing. From a risk point of view, coastal erosion is the most widespread and continuous process. However, catastrophic events play a very significant role (tsunami, hurricane and storm surge), both for coastal erosion and human damage. In Europe, coastal erosion is widespread (at least 70%) distributed very irregularly.

## EROSION AND ACCRETION

### Extraordinary Processes

#### *Tsunamis*

Tsunamis, potentially enormous waves often caused by earthquakes, have great erosional and sediment-reworking potential. They may strip beaches of sand that may have taken years to accumulate and may destroy trees and other coastal vegetation. Tsunamis are also capable of flooding hundreds of meters inland past the typical high-water level and fast-moving water, associated with the inundating tsunami, can crush homes and other coastal structures.

### Storm Surges and Hurricanes

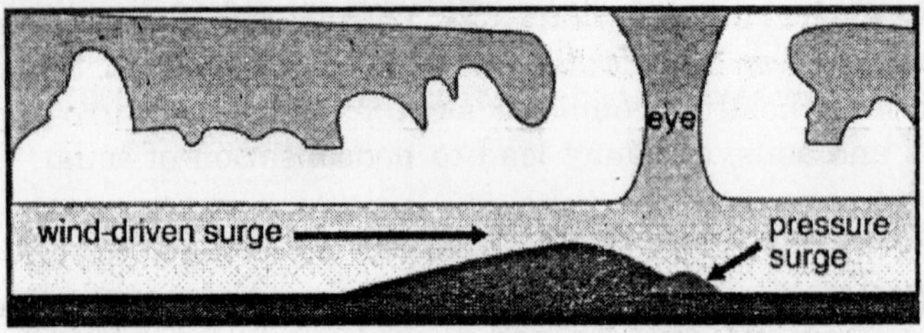


A storm surge is an onshore gush of water associated with a low pressure weather system. The most extreme storm surges result from extreme weather systems, such as tropical cyclones or hurricanes, but storm surges can also be produced by less powerful storms. Storm surges can cause beach accretion and erosion. Historically notable storm surges occurred during the North Sea Flood of 1953, Hurricane Katrina, and the 1970 Bhola cyclone.

### Gradual Processes

The gradual evolution of beaches often comes from the interaction of longshore drift, a waven-driven process by which sediments move along a beach shore, and other sources of erosion or accretion, such as nearby rivers.

### Deltas (Rhône, Ebro, Nile)

Deltas are nourished by alluvial systems and accumulate sand and silt, growing where the sediment flux from land is large enough to avoid complete removal by coastal currents, tides, or waves. Most modern deltas formed during the last five thousand years, after the present sea-level high stand was attained. However, not all sediment remains permanently in place: in the short term (decades to centuries), exceptional river floods, storms or other energetic events may remove significant portions of delta sediment or change its lobe distribution and, on longer geological time scales, sea-level fluctuations lead to destruction of deltaic features.

## Historical Accretion of European Beaches

In the Mediterranean sea, deltas have been continuously growing during for the last several thousand years. Six to seven thousand years ago, the sea level stabilized, and continuous river systems, ephemeral torrents, and other factors began this steady accretion. Since intense human use of coastal areas is a relatively recent phenomenon (except in the Nile delta), beach contours were primarily shaped by natural forces until the last centuries.

In Barcelona, for example, the accretion of the coast was a natural process until the late middle ages. At that time, the initiation of harbour-building increased the rate of accretion.

The port of Ephesus, one of the great cities of the Ionian Greeks in Asia Minor, was filled with sediment due to accretion from a nearby river; it is now 5km from the sea. Likewise, Ostia, the once-important port near ancient Rome, is now several kilometers inland, the coastline having moved slowly seaward.

Bruges, a port during the early middle ages, was accessible from the sea until around 1050. At that time, however, the natural link between Bruges and the sea silted up. In 1134, a storm flood caused a deep channel, the Zwin, to appear, and the city remained linked to the sea until the fifteenth century via a canal from the Zwin to Bruges. But Bruges had to use a number of outports, such as Damme and Sluis, for this purpose. In 1907, a new sea-port was inaugurated in Zeebrugge.

## Modern Recession of Beaches

At the present important segment of low coasts are in recession, losing sand and reducing the beaches' dimensions.

This loss could occur very rapidly. From more to less natural causes (degree of anthropization downwards):

**Sete:** The coast recession near Sete is related with coastal drift sand supply interruption due to growth of the Rhone delta, which (like most deltas) is becoming independent of the rest of the coast. The present lido shoreline is 210 meters away from the Roman lido.

### California Beaches

California's beaches and other shoreline features change according to the availability of beach sand, the wave and current energy impinging on the coast, and other physical processes that affect the movement of sand.

A constant supply of sand is necessary for beaches to form and be maintained along this shoreline. Many human activities, including dam construction and river channelization, have reduced the supply of sand that reaches the ocean.

This, in turn, has prevented beaches from being replenished and has thus created greater vulnerability for shorelines that have always been subject to varying levels of erosion.

There are few practical solutions to improving sand supply from inland sources, so management of shoreline erosion will likely continue to focus at the land/sea interface along the California coastline.

Construction of breakwaters, jetties, or groyne fields to protect harbor entrances, maintain beaches, or protect coastal structures have both helped and harmed the movement of sand along the shoreline. Protective armoring formations trap sand and allow beaches to expand up-coast from the device, but can interrupt the flow of sand to beaches located down-coast.

### Poland

During the last glaciation, the Baltic Polish area was covered in ice and associated morainal sediments. Deglaciation left a substantial amount of unconsolidated sediment. Currently, these unconsolidated sediments are strongly eroded and reworked by the sea.

### Aveiro

The North Portuguese coast and its beaches were fed by large Iberian rivers. The massive building of dams in the Douro River basin has cut the sediment supply to the Aveiro coast, resulting in its recession. Hard protective works have been done all along.

### Holland

The Dutch coast consists of sandy, multi-barred beaches and can be characterised as a wave-dominated coast. Approximately 290 km of the coast consists of dunes and 60 km is protected by structures such as dikes and dams. With the melting of the ice at the end of the last ice age the coastline shifted eastward until about 5000 years ago the present position of the Dutch coastline was reached.

As the sea level rise stagnated, the sand supply decreased and the formation of the beach ridges stopped, after which when the sea broke through the lines of dunes during storms, men started to defend the land by building primitive dikes and walls. The dunes, together with the beach and the shoreline, offer a natural, sandy defence to the sea. About 30% of the Netherlands lies below sea level.

Over the last 30 years, approximately 1 million $m^3$ sand per year has been lost from the Dutch coast to deep water. In most northern coastal sections, erosion occurs in deep water and also in the nearshore zone. In most southern sections, sedimentation occurs in the nearshore zone and erosion in deep water. Structural erosion is due to sea level rise relative to the land and, in some spots, it is caused by harbour dams. The Dutch coast looked at as a single unit shows erosive behaviour. Approximately 12 million m3 of sand is transferred annually from the North Sea to the Wadden Sea as a result of relative rising sea level and coastal erosion.

## RELATIVE SEA LEVEL CHANGES

Several geological events and the climate can change (progressively or suddenly) the relative height of the Earth's surface to the sea-level. The coastline is continuouly changing by these events or processes.

## EXTRAORDINARY PROCESSES

### Earthquakes

Some earthquakes can create sudden variations of relative ground level and change the coastline dramatically. Structurally controlled coasts include the San Andreas fault zone in California and the seismic Mediterranean belt (from Gibraltar to Greece).

The tremors, which reached a peak on October 4, 1983, damaged 8,000 buildings in the city center and raised the sea bottom by almost 2m. This rendered the Bay of Pozzuoli too shallow for large craft and required the reconstruction of the harbour with new quays. This photo shows the new quay.

### Volcanism

Volcanic activity can create new islands. Surtsey Island, Iceland, for example, was created between November of 1963 and June of 1967. The 800m-diameter island has since been partially eroded by waves, rain, and wind, but it is expected to last another 100 years.

### Gradual Processes: Subsidence and Uplift

Subsidence is the motion of the Earth's surface downward relative to the sea level due to internal geodynamic causes. The opposite of subsidence is uplift, which results in an increase in elevation.

Venice is probably the best-known example of a subsiding location. It experiences periodic flooding when tides or surges arrive. This phenomenon is caused by the compaction of young sediments in the Po river delta area, magnified by water and gas subsurface exploitation. Man-made works to solve this progressive sinking have been unsuccessful.

Mälaren, the third-largest lake in Sweden, is an example of deglacial uplift. It was once a bay on which seagoing vessels were once able to sail far into the interior of Sweden, but it ultimately became a lake. Its uplift was caused by deglaciation: the removal of the weight of ice-age glaciers caused rapid uplift of the depressed land. For 2,000 years as the ice was unloaded, uplift proceeded at about 7.5 cm/year. Once deglaciation was complete, uplift slowed to about 2.5 cm/year, and it decreased exponentially after that. Today, typical uplift rates are 1 cm/year or less, and studies suggest that rebound will continue for about another 10,000 years. The total uplift from the end of deglaciation may be up to 400 m.

## Integrated Coastal Zone Management

Integrated coastal zone management (ICZM) is a process for the management of the coast using an integrated approach, regarding all aspects of the coastal zone, including geographical and political boundaries, in an attempt to achieve sustainability.

This concept was borne in 1992 during the Earth Summit of Rio de Janeiro. The policy regarding ICZM is set out in the proceedings of the summit within Agenda 21, Chapter 17.

The European Commission defines the ICZM as follows:-

*ICZM is a dynamic, multidisciplinary and iterative process to promote sustainable management of coastal zones. It covers the full cycle of information collection, planning (in its broadest sense), decision making, management and monitoring of implementation. ICZM uses the informed participation and cooperation of all stakeholders to assess the societal goals in a given coastal area, and to take actions towards meeting these objectives. ICZM seeks, over the long-term, to balance environmental, economic, social, cultural and recreational objectives, all within the limits set by natural dynamics. 'Integrated' in ICZM refers to the integration of objectives and also to the integration of the many instruments needed to meet these objectives. It means integration of all relevant policy areas, sectors, and levels of administration. It means integration of the terrestrial and marine components of the target territory, in both time and space.*

## Environmental Geography

Environmental geography is the branch of geography that describes the spatial aspects of interactions between humans and the natural world. It requires an understanding of the dynamics of geology, meteorology, hydrology, biogeography, and geomorphology, as well as the ways in which human societies conceptualize the environment. The links between cultural and physical geography were once more readily apparent than they are today. As human experience of the world is increasingly mediated by technology, the relationships have often become obscured.

Environmental geography represents a critically important set of analytical tools for assessing the impact of human presence on the environment by measuring the result of human activity on natural landforms and cycles.

## Environmental Management

Environmental Management is not, as the phrase suggests, the management of the environment as such but rather the management of the humankind's interaction with and impact upon the environment. The three main issues that affect managers are issues involving politics (networking), programs (projects), and resources (i.e. money, facilities, ect). The need for environmental management can be viewed from a variety of perspectives. A more common philosophy and impetus behind environmental management is the concept of carrying capacity. Simply put, carrying capacity refers to the maximum number of organisms a particular resource can sustain. The concept of carrying capacity, whilst understood by many cultures over history, has its roots in Malthusian theory. A common example of the consequences of exceeding the carrying capacity of an area is the starvation and eventual cannibalism of tribes on the Easter Islands after the depletion of the island's resources. Environmental management is therefore not the conservation of the environment solely for the environment's sake, but rather the conservation of the environment for humankind's sake. This element of sustainable exploitation, getting the most out of natural assets, is visible in the French approach to water resources.

Environmental management involves the management of all components of the bio-physical environment, both living (biotic) and non-living (abiotic). This is due to the interconnected and network of relationships amongst all living species and their habitats. The environment also involves the relationships of the human environment, such as the social, cultural and economic environment with the bio-physical environment. As with all management functions, effective management tools, standards and systems are required. An environmental management standard or system or protocol attempts to reduce environmental impact as measured by some objective criteria. The ISO 14001 standard is the most widely used standard for environmental risk management and is closely aligned to the European Eco Management & Audit Scheme (EMAS). As a common auditing standard, the ISO 19011 standard explains how to combine this with quality management. The UK has developed a phased standard (BS8555) that can help smaller companies move to ISO 14001 in six manageable steps.

Other environmental management systems tend to be based on this standard and to extend it in various ways:

- The Natural Step focuses on basic sustainability criteria and helps focus engineering on reducing use of materials or energy use that is unsustainable in the long term
- Natural Capitalism advises using accounting reform and a general biomimicry and industrial ecology approach to do the same thing
- US Environmental Protection Agency has many further terms and standards that it defines as appropriate to large-scale EMS
- The UN and World Bank has encouraged adopting a "natural capital" measurement and management framework.
- The European Union Eco-Management and Audit Scheme (EMAS)

Other strategies exist that rely on making simple distinctions rather than building top-down management "systems" using performance audits and full cost accounting. For instance, Ecological Intelligent Design divides products into consumables, service products or durables and unsaleables - toxic products that no one should buy, or in many cases, do not realize they are buying. By eliminating the unsaleables from the comprehensive outcome of any purchase, better environmental management is achieved without "systems".

## GEODESY

Geodesy, also called geodetics, is the scientific discipline that deals with the measurement and representation of the earth, its gravitational field, and other geodynamic phenomena, such as crustal motion, oceanic tides, and polar motion.

### Definition

Geodesy is primarily concerned with positioning and the gravity field and geometrical aspects of their temporal variations, although it can also include the study of the Earth's magnetic field. Especially in the German speaking world, geodesy is divided in geomensuration ("Erdmessung" or "höhere Geodäsie"), which is concerned with measuring the earth on a global scale, and surveying ("Ingenieurgeodäsie"), which is concerned with measuring parts of the surface.

The shape of the earth is to a large extent the result of its rotation, which causes its equatorial bulge, and the competition of geologic processes such as the collision of plates and of vulcanism, resisted by the earth's gravity field. This applies to the solid surface, the liquid surface (dynamic sea surface topography) and the earth's atmosphere. For this reason, the study of the Earth's gravity field is seen as a part of geodesy, called physical geodesy.

### History

### Geoid and Reference Ellipsoid

The geoid is essentially the figure of the Earth abstracted from its topographic features. It is an idealized equilibrium surface of sea water, the mean sea level surface in the absence of currents, air pressure variations etc. and continued under the continental masses. The geoid, unlike the ellipsoid, is irregular and too complicated to serve as the computational surface on which to solve geometrical problems like point positioning. The geometrical separation between it and the reference ellipsoid is called the geoidal undulation. It varies globally between ±110 m. A reference ellipsoid, customarily chosen to be the same size (volume) as the geoid, is described by its semi-major axis (equatorial radius) a and flattening f. The quantity f = (a-b)/a, where b is the semi-minor axis (polar radius), is a purely geometrical one. The mechanical ellipticity of the earth (dynamical flattening, symbol J2) is determined to high precision by observation of satellite orbit perturbations. Its relationship with the geometric flattening is indirect. The relationship depends on the internal density distribution, or, in simplest terms, the degree of central concentration of mass.

The 1980 Geodetic Reference System (GRS80) posited a 6,378,137 m semi-major axis and a 1:298.257 flattening. This system was adopted at the XVII General Assembly of the International Union of Geodesy and Geophysics (IUGG). It is essentially the basis for geodetic positioning by the Global Positioning System and is thus also in extremely widespread use outside the geodetic community.

The numerous other systems which have been used by diverse countries for their maps and charts are gradually dropping out of use as more and more countries move to global, geocentric reference systems using the GRS80 reference ellipsoid.

## Coordinate Systems in Space

The locations of points in three-dimensional space are most conveniently described by three cartesian or rectangular coordinates, X,Y and Z. Since the advent of satellite positioning, such coordinate systems are typically geocentric: the Z axis is aligned with the Earth's (conventional or instantaneous) rotation axis.

Before the satellite geodesy era, the coordinate systems associated with geodetic datums attempted to be geocentric, but their origins differed from the geocentre by hundreds of metres, due to regional deviations in the direction of the plumbline (vertical). These regional geodetic datums, such as ED50 (European Datum 1950) or NAD83 (North American Datum 1983) have ellipsoids associated with them that are regional 'best fits' to the geoids within their areas of validity, minimising the deflections of the vertical over these areas.

It is only because GPS satellites orbit about the geocentre, that this point becomes naturally the origin of a coordinate system defined by satellite geodetic means, as the satellite positions in space are themselves computed in such a system. Geocentric coordinate systems used in geodesy can be divided naturally into two classes:

1. Inertial reference systems, where the coordinate axes retain their orientation relative to the fixed stars, or equivalently, to the rotation axes of ideal gyroscopes; the X axis points to the vernal equinox
2. Co-rotating, also ECEF ("Earth Centred, Earth Fixed"), where the axes are attached to the solid body of the Earth. The X axis lies within the Greenwich observatory's meridian plane.

The coordinate transformation between these two systems is described to good approximation by (apparent) sidereal time, which takes into account variations in the Earth's axial rotation (length-of-day variations). A more accurate description also takes polar motion into account, a phenomenon currently closely monitored by geodesists.

**Coordinate Systems in the Plane**

In surveying and mapping, important fields of application of geodesy, two general types of coordinate systems are used in the plane:

1. Plano-polar, in which points in a plane are defined by a distance s from a specified point along a ray having a specified direction a with respect to a base line or axis;
2. Rectangular, points are defined by distances from two perpendicular axes called x and y. It is geodetic practice - contrary to the mathematical convention - to let the x axis point to the North and the y axis to the East.

Rectangular coordinates in the plane can be used intuitively with respect to one's current location, in which case the x axis will point to the local North. More formally, such coordinates can be obtained from three-dimensional coordinates using the artifice of a map projection. It is not possible to map the curved surface of the Earth onto a flat map surface without deformation. The compromise most often chosen - called a conformal projection- preserves angles and length ratios, so that small circles are mapped as small circles and small squares as squares.

An example of such a projection is UTM (Universal Transverse Mercator). Within the map plane, we have rectangular coordinates x and y. In this case the North direction used for reference is the map North, not the local North. The difference between the two is called meridian convergence.

It is easy enough to "translate" between polar and rectangular coordinates in the plane: let, as above, direction and distance be a and s respectively, then we have

The reverse translation is slightly more tricky.

## Heights

In geodesy, point or terrain heights are "above sea level", an irregular, physically defined surface. Therefore a height should ideally not be referred to as a coordinate. It is more like a physical quantity, and though it can be tempting to treat height as the vertical coordinate z, in addition to the horizontal coordinates x and y, and though this actually is a good approximation of physical reality in small areas, it quickly becomes invalid in larger areas.

Heights come in the following variants:

1. Orthometric heights
2. Normal heights
3. Geopotential heights

Each has its advantages and disadvantages. Both orthometric and normal heights are heights in metres above sea level, whereas geopotential numbers are measures of potential energy (unit: m2s - 2) and not metric. Orthometric and normal heights differ in the precise way in which mean sea level is conceptually continued under the continental masses. The reference surface for orthometric heights is the geoid, an equipotential surface approximating mean sea level.

None of these heights is in any-way related to geodetic or ellipsoidial heights, which express the height of a point above the reference ellipsoid. Satellite positioning receivers typically provide ellipsoidal heights, unless they are fitted with special conversion software based on a model of the geoid.

## Geodetic Data

Because geodetic point coordinates (and heights) are always obtained in a system that has been constructed itself using real observations, we have to introduce the concept of a geodetic datum: a physical realization of a coordinate system used for describing point locations. The realization is the result of choosing conventional coordinate values for one or more datum points.

In the case of height datums, it suffices to choose one datum point: the reference bench mark, typically a tide gauge at the shore. Thus we have vertical datums like the NAP (Normaal Amsterdams Peil), the North American Vertical Datum 1988 (NAVD88), the Kronstadt datum, the Trieste datum, etc.

In case of plane or spatial coordinates, we typically need several datum points. A regional, ellipsoidal datum like ED50 can be fixed by prescribing the undulation of the geoid and the deflection of the vertical in one datum point, in this case the Helmert Tower in Potsdam. However, an overdetermined ensemble of datum points can also be used.

Changing the coordinates of a point set referring to one datum, to make them refer to another datum, is called a datum transformation. In the case of vertical datums, this consists of simply adding a constant shift to all height values. In the case of plane or spatial coordinates, datum transformation takes the form of a similarity or Helmert transformation, consisting of a rotation and scaling operation in addition to a simple translation. In the plane, a Helmert transformation has four parameters, in space, seven.

## A Note on Terminology

In the abstract, a coordinate system as used in mathematics and geodesy is, e.g., in ISO terminology, referred to as a coordinate system. International geodetic organizations like the IERS (International Earth Rotation and Reference Systems Service) speak of a reference system.

When these coordinates are realized by choosing datum points and fixing a geodetic datum, ISO uses the terminology coordinate reference system, while IERS speaks of a reference frame. A datum transformation again is referred to by ISO as a coordinate transformation. (ISO 19111: Spatial referencing by coordinates).

## Point Positioning

Point positioning is the determination of the coordinates of a point on land, at sea, or in space with respect to a coordinate system. Point position is solved by computation from measurements linking the known positions of terrestrial or extraterrestrial points with the unknown terrestrial position. This may involve transformations between or among astronomical and terrestrial coordinate systems.

The known points used for point positioning can be triangulation points of a higher order network, or GPS satellites.

Traditionally, a hierarchy of networks has been built to allow point positioning within a country. Highest in the hierarchy were triangulation networks. These were densified into networks of traverses (polygons), into which local mapping surveying measurements, usually with measuring tape, corner prism and the familiar red and white poles, are tied.

Nowadays all but special measurements (e.g., underground or high precision engineering measurements) are performed with GPS. The higher order networks are measured with static GPS, using differential measurement to determine vectors between terrestrial points. These vectors are then adjusted in traditional network fashion. A global polyhedron of permanently operating GPS stations under the auspices of the IERS is used to define a single global, geocentric reference frame which serves as the "zero order" global reference to which national measurements are attached. For surveying mappings, frequently Real Time Kinematic GPS is employed, tying in the unknown points with known terrestrial points close by in real time.

One purpose of point positioning is the provision of known points for mapping measurements, also known as (horizontal and vertical) control. In every country, thousands of such known points exist in the terrain and are documented by the national mapping agencies. Constructors and surveyors involved in real estate will use these to tie their local measurements to.

## Geodetic Problems

In geometric geodesy, two standard problems exist:

*Geodetic principal problem (also: first geodetic problem)*

*Given a point (in terms of its coordinates) and the direction (azimuth) and distance from that point to a second point, determine (the coordinates of) that second point.*

*Geodetic inverse problem (also: second geodetic problem)*

*Given two points, determine the azimuth and length of the line (straight line, arc or geodesic) that connects them.*

In the case of plane geometry (valid for small areas on the Earth's surface) the solutions to both problems reduce to simple trigonometry. On the sphere, the solution is significantly more complex, e.g., in the inverse problem the azimuths will differ between the two end points of the connecting great circle, arc, i.e. the geodesic.

On the ellipsoid of revolution, solutions in closed form do not exist, so rapidly converging series expansions have traditionally been used, such as Vincenty's formulae.

In the general case, the solution is called the geodesic for the surface considered. It may be nonexistent or non-unique. The differential equations for the geodesic can be solved numerically, e.g., in MATLAB.

## Geodetic Observational Concepts

Here we define some basic observational concepts, like angles and coordinates, defined in geodesy (and astronomy as well), mostly from the viewpoint of the local observer.

- The plumbline or vertical is the direction of local gravity, or the line that results by following it. It is slightly curved.
- The zenith is the point on the celestial sphere where the direction of the gravity vector in a point, extended upwards, intersects it. More correct is to call it a <direction> rather than a point.

- The nadir is the opposite point (or rather, direction), where the direction of gravity extended downward intersects the (invisible) celestial sphere.
- The celestial horizon is a plane perpendicular to a point's gravity vector.
- Azimuth is the direction angle within the plane of the horizon, typically counted clockwise from the North (in geodesy and astronomy) or South (in France).
- Elevation is the angular height of an object above the horizon, Alternatively zenith distance, being equal to 90 degrees minus elevation.
- Local topocentric coordinates are azimuth (direction angle within the plane of the horizon) and elevation angle (or zenith angle) and distance.
- The North celestial pole is the extension of the Earth's (precessing and nutating) instantaneous spin axis extended Northward to intersect the celestial sphere. (Similarly for the South celestial pole.)
- The celestial equator is the intersection of the (instantaneous) Earth equatorial plane with the celestial sphere.
- A meridian plane is any plane perpendicular to the celestial equator and containing the celestial poles.
- The local meridian is the plane containing the direction to the zenith and the direction to the celestial pole.

## Geodetic Measurements

The level is used for determining height differences and height reference systems, commonly referred to mean sea level. The traditional spirit level produces these practically most useful heights above sea level directly; the more economical use of GPS instruments for height determination requires precise knowledge of the figure of the geoid, as GPS only gives heights above the GRS80 reference ellipsoid. As geoid knowledge accumulates, one may expect use of GPS heighting to spread.

The theodolite is used to measure horizontal and vertical angles to target points. These angles are referred to the local vertical. The tacheometer additionally determines, electronically or electro-optically, the distance to target, and is highly automated in its operations. The method of free station position is widely used.

For local detail surveys, tacheometers are commonly employed although the old-fashioned rectangular technique using angle prism and steel tape is still an inexpensive alternative. More and more, also real time kinematic (RTK) GPS techniques are used. Data collected is tagged and recorded digitally for entry into a Geographic Information System (GIS) data base.

Geodetic GPS receivers produce directly three-dimensional coordinates in a geocentric coordinate frame. Such a frame is, e.g., WGS84, or the frames that are regularly produced and published by the International Earth Rotation and Reference Systems Service (IERS).

GPS receivers have almost completely replaced terrestrial instruments for large-scale base network surveys. For planet-wide geodetic surveys, previously impossible, we can still mention Satellite Laser Ranging (SLR) and Lunar Laser Ranging (LLR) and Very Long Baseline Interferometry (VLBI) techniques. All these techniques also serve to monitor Earth rotation irregularities as well as plate tectonic motions.

Gravity is measured using gravimeters. There are two basic kinds of gravimeters. Absolute gravimeters, which nowadays can also be used in the field, are based directly on measuring the acceleration of free fall (for example, of a reflecting prism in a vacuum tube). They are used for establishing the vertical geospatial control. Most common relative gravimeters are spring based.

They are used in gravity surveys over large areas for establishing the figure of the geoid over these areas. Most accurate relative gravimeters are superconducting gravimeters, and these are sensitive to one thousandth of one billionth of the Earth surface gravity. Twenty-some superconducting gravimeters are used worldwide for studying Earth tides, rotation, interior, and ocean and atmospheric loading, as well as for verifying the Newtonian constant of gravitation.

## Units and Measures on the Ellipsoid

Geographical latitude and longitude are stated in the units degree, minute of arc, and second of arc. They are angles, not metric measures, and describe the direction of the local normal to the reference ellipsoid of revolution. This is approximately the same as the direction of the plumbline, i.e., local gravity, which is also the normal to the geoid surface. For this reason, astronomical position determination, measuring the direction of the plumbline by astronomical means, works fairly well provided an ellipsoidal model of the figure of the Earth is used.

A geographic mile, defined as one minute of arc on the equator, equals 1,855.32571922 m. A nautical mile is one minute of astronomical latitude. The radius of curvature of the ellipsoid varies with latitude, being the longest at the pole and shortest at the equator as is the nautical mile.

A metre was originally defined as the 40 millionth part of the length of a meridian. This means that a kilometre is equal to (1/40,000) * 360 * 60 meridional minutes of arc, which equals 0.54 nautical miles. Similarly a nautical mile is on average 1/0.54 = 1.85185... km.

## Temporal Change

In geodesy, temporal change can be studied by a variety of techniques. Points on the Earth's surface change their location due to a variety of mechanisms:

- Continental plate motion, plate tectonics
- Episodic motion of tectonic origin, esp. close to fault lines
- Periodic effects due to Earth tides
- Postglacial land uplift due to isostatic adjustment
- Various anthropogenic movements due to, e.g., petroleum or water extraction or reservoir construction.

The science of studying deformations and motions in the Earth's crust and the solid Earth as a whole is called geodynamics. Often, also study of the Earth's irregular rotation is included in its definition.

Techniques for studying geodynamic phenomena on the global scale include:

- satellite positioning by GPS and similar techniques,
- very long baseline interferometry (VLBI)
- satellite and lunar laser ranging.
- Regionally and locally, precise levelling,
- precise tacheometers
- monitoring gravity change,
- Synthetic Aperture Radar Interferometry (InSAR) using satellite images.

# 3

# GEOMORPHOLOGY

Geomorphology (from Greek: ge, "earth"; morfé, "form"; and, logos, "knowledge") is the study of landforms, including their origin and evolution, and the processes that shape them. The underlying question is: Why do landscapes look the way they do? Geomorphologists seek to understand landform history and dynamics, and predict future changes through a combination of field observation, physical experiment, and numerical modeling. The discipline is practiced within geology, geodesy, geography, archaeology, and civil and environmental engineering. Early studies in geomorphology are the foundation for pedology, one of two main branches of soil science.

Landforms evolve in response to a combination of natural and anthropogenic processes. The landscape is built up through tectonic uplift and volcanism. Denudation occurs by erosion and mass wasting, which produces sediment that is transported and deposited elsewhere within the landscape or off the coast. Landscapes are also lowered by subsidence, either due to tectonics or physical changes in underlying sedimentary deposits. These processes are each influenced differently by climate, ecology, and human activity.

Practical applications of geomorphology include measuring the effects of climate change, hazard assessments including landslide prediction and mitigation, river control and restoration, coastal protection, and assessing the presence of water on Mars.

Paleogeomorphology is the study of the geomorphology of all or part of the earth's surface at some time in the earth's past.

## HISTORY

Perhaps the earliest one to devise a theory of geomorphology was the polymath Chinese scientist and statesman Shen Kuo (1031-1095 AD). This was based on his observation of marine fossil shells in a geological stratum of a mountain hundreds of miles from the Pacific Ocean.

Noticing bivalve shells running in a horizontal span along the cut section of a cliffside, he theorized that the cliff was once the pre-historic location of a seashore that had shifted hundreds of miles over the centuries. He inferred that the land was reshaped and formed by soil erosion of the mountains and by deposition of silt, after observing strange natural erosions of the Taihang Mountains and the Yandang Mountain near Wenzhou. Furthermore, he promoted the theory of gradual climate change over centuries of time once ancient petrified bamboos were found to be preserved underground in the dry, northern climate zone of Yanzhou, which is now modern day Yan'an, Shaanxi province.

Geomorphology was not originally differentiated from the rest of geography. The first geomorphic model was the geographical cycle or the cycle of erosion, developed by William Morris Davis between 1884 and 1899. The cycle was inspired by theories of uniformitarianism which were first formulated by James Hutton (1726-1797). Concerning valley forms, the cycle was depicted as a sequence by which a river would cut a valley more and more deeply, but then erosion of side valleys would eventually flatten out the terrain again, now at a lower elevation. The cycle could be started over by uplift of the terrain. The model is today considered too much of a simplification to be especially useful in practice.

Walther Penck developed an alternative model in the 1920s, based on ratios of uplift and erosion, but it was also too weak to explain a variety of landforms. G. K. Gilbert was an important early American geomorphologist.

## PROCESSES

Modern geomorphology focuses on the quantitative analysis of interconnected processes, such as the contribution of solar energy, the rates of steps of the hydrologic cycle, and plate movement rates from geophysics to compute the age and expected fate of landforms. The use of more precise measurement technique has also enabled processes like erosion to be observed directly, rather than merely surmised from other evidence. Computer simulation is also valuable for testing that a particular model yields results with properties similar to real terrain.

Primary surface processes responsible for most topographic features include wind, waves, weathering, mass wasting, ground water, surface water, glaciers, tectonism, and volcanism.

## Fluvial Geomorphology

Rivers and streams are not only conduits of water, but also of sediment. The water, as it flows over the channel bed, is able to mobilize sediment and transport it downstream, either as bedload, suspended load or dissolved load. The rate of sediment transport depends on the availability of sediment itself and on the river's discharge.

As rivers flow across the landscape, they generally increase in size, merging with other rivers. The network of rivers thus formed is a drainage system and is often dendritic, but may adopt other patterns depending on the regional topography and underlying geology.

## Glacial Geomorphology

Glaciers, while geographically restricted, are effective agents of landscape change. The gradual movement of ice down a valley causes abrasion and plucking of the underlying rock. Abrasion produces fine sediment, termed glacial flour. The debris transported by the glacier, when the glacier recedes, is termed a moraine. Glacial erosion is responsible for U-shaped valleys, as opposed to the V-shaped valleys of fluvial origin.

## Weathering

This results from chemical dissolution of rock and from the mechanical wearing of rock by plant roots, ice expansion, and the abrasive action of sediment. Weathering provides the source of the sediment transported by fluvial, glacial, aeolian, or biotic processes.

## TAXONOMY

Different geomorphological processes dominate at different spatial and temporal scales. To help categorize landscape scales some geomorphologists use the following taxonomy:

- 1st - Continent, ocean basin, climatic zone (~10,000,000 km$^2$)
- 2nd - Shield, e.g. Baltic shield, or mountain range (~1,000,000 km$^2$)
- 3rd - Isolated sea, Sahel (~100,000 km$^2$)
- 4th - Massif, e.g. Massif Central or Group of related landforms, e.g., Weald (~10,000 km$^2$)
- 5th - River valley, Cotswolds (~1,000 km$^2$)
- 6th - Individual mountain or volcano, small valleys (~100 km$^2$)
- 7th - Hillslopes, stream channels, estuary (~10 km$^2$)
- 8th - gully, barchannel (~1 km$^2$)
- 9th - Meter-sized features

## GLACIOLOGY

Glaciology is the study of glaciers, or more generally ice and natural phenomena that involve ice. The word glacier is derived from the Latin glaciees, meaning ice or frost.

Glaciology is an interdisciplinary earth science that integrates geophysics, geology, physical geography, geomorphology, climatology, meteorology, hydrology, biology, and ecology. The impact of glaciers on humans adds the fields of human geography and anthropology. The presence of ice on Mars and Europa brings in an extraterrestrial component to the field.

### Overview

Areas of study within glaciology include glacial history and the reconstruction of past glaciation patterns, effects of glaciers on climate and vice versa, the dynamics of ice movement, the contributions of glaciers to erosion and geomorphology, and lifeforms that live in the ice. A Glaciologist is a person who studies glaciers. Glaciology is one of the key areas of polar research.

## Types

There are two general categories of glaciation which glaciologists distinguish: alpine glaciation, accumulations or "rivers of ice" confined to valleys; and continental glaciation, unrestricted accumulations which once covered much of the northern continents.

- Alpine - ice flows down the valleys of mountainous areas and forms a tongue of ice moving towards the plains below. Alpine glaciers tend to make the topography more rugged.
- Continental - an ice sheet found today, only in high latitudes (Greenland/Antarctica), thousands of square kilometers wide and thousands of meters thick. These tend to smooth out the landscape.

## Zones of Glaciers

- Accumulation, where the formation of ice is faster than its removal.
- Wastage or Ablation, where the sum of melting and evaporation (sublimation) is greater than the amount of snow added each year.

## Movement

### *Ablation*

wastage through evaporation and melting

### *Arête*

an acute ridge of rock where two cirques abut.

### *Bergshrund*

crevasse formed near the head of a glacier, where the mass of ice has rotated, sheared and torn itself apart in the manner of a geological fault.

*Cirque, Corrie or Cwm*

bowl shaped depression excavated by the source of a glacier.

*Creep*

adjustment to stress at a molecular level.

*Flow*

movement (of ice) in a constant direction.

*Fracture*

brittle failure (breaking of ice) under the stress raised when movement is too rapid to be accommodated by creep. It happens for example, as the central part of a glacier movinges faster than the edges.

*Horn*

spire of rock formed by the headward erosion of a ring of cirques around a single mountain. It is an extreme case of an arête.

*Plucking/Quarrying*

where the adhesion of the ice to the rock is stronger than the cohesion of the rock, part of the rock leaves with the flowing ice.

*Tarn*

a lake formed in the bottom of a cirque when its glacier has melted.

*Tunnel Valley*

The tunnel is that formed by hydraulic erosion of ice and rock below an ice sheet margin. The tunnel valley is what remains of it in the underlying rock when the ice sheet has melted.

**Glacial Deposits**

***Stratified***

Outwash sand/gravel

from front of glaciers, found on a plain

*Kettles*

block of stagnant ice leaves a depression or pit

*Eskers*

steep sided ridges of gravel/sand, possibly caused by streams running under stagnant ice

*Kames*

stratified drift builds up low steep hills

*Varves*

alternating thin sedimentary beds (coarse and fine) of a proglacial lake. Summer conditions deposit more and coarser material and those of the winter, less and finer.

*Unstratified*

Till-unsorted

(glacial flour to boulders) deposited by receding/advancing glaciers, forming moraines, and drumlins

*Moraines*

(Terminal) material deposited at the end; (Ground) material deposited as glacier melts; (lateral) material deposited along the sides.

*Drumlins*

smooth elongated hills composed of till.

*Ribbed Moraines*

large subglacial elongated hills transverse to former ice flow.

## HYDROLOGY

Hydrology (from Greek: Yd, hudor, "water"; and, logos, "knowledge") is the study of the movement, distribution, and quality of water throughout the Earth, and thus addresses both the hydrologic cycle and water resources. A practitioner of hydrology is a hydrologist, working within the fields of either earth or environmental science, physical geography or civil and environmental engineering.

Domains of hydrology include hydrometeorology, surface hydrology, hydrogeology, drainage basin management and water quality, where water plays the central role. Oceanography and meteorology are not included because water is only one of many important aspects.

Hydrological research is useful in that it allows us to better understand the world in which we live, and also provides insight for environmental engineering, policy and planning.

### History of Hydrology

Hydrology has been a subject of investigation and engineering for millennia. For example, in about 4000 B.C. the Nile was dammed to improve agricultural productivity of previously barren lands. Mesopotamia/Mesopotamian towns were protected from flooding with high earthen walls. Aqueducts were built by the Ancient Greece/Greeks and Ancient Romans, while the History of China built irrigation and flood control works. The ancient Sinhalese used hydrology to build complex Irrigation Works of Ancient Sri Lanka, known for invention of the Valve Pit which allowed construction of large reservoirs, anicuts and canals which still function.

Vitruvius/Marcus Vitruvius, in the first century B.C., described a philosophical theory of the hydrologic cycle, in which precipitation falling in the mountains infiltrated the earth's surface and led to streams and springs in the lowlands. With adoption of a more scientific approach, Leonardo da Vinci and Bernard Palissy independently reached an accurate representation of the hydrologic cycle. It was not until the 17th century that hydrologic variables began to be quantified.

Pioneers of the modern science of hydrology include Pierre Perrault, Edme Mariotte and Edmund Halley. By measuring rainfall, runoff, and drainage area, Perrault showed that rainfall was sufficient to account for flow of the Seine. Marriotte combined velocity and river cross-section measurements to obtain discharge, again in the Seine. Halley showed that the evaporation from the Mediterranean Sea was sufficient to account for the outflow of rivers flowing into the sea.

Advances in the 18th century included the Bernoulli piezometer and Bernoulli's equation, by Daniel Bernoulli, the Pilot tube. The 19th century saw development in groundwater hydrology, including Darcy's law, the Dupuit-Thiem well formula, and Hagen-Poiseuille's capillary flow equation.

Rational analyses began to replace empiricism in the 20th century, while governmental agencies began their own hydrological research programs. Of particular importance were Leroy Sherman's unit hydrograph, the infiltration theory of Robert E. Horton, and C.V. Theis's Aquifer test/equation describing well hydraulics. Since the 1950's, hydrology has been approached with a more theoretical basis than in the past, facilitated by advances in the physical understanding of hydrological processes and by the advent of computers and especially Geographic Information Systems (GIS).

## Hydrologic Cycle

The central theme of hydrology is that water moves throughout the Earth through different pathways and at different rates. The most vivid image of this is in the evaporation of water from the ocean, which forms clouds. These clouds drift over the land and produce rain. The rainwater flows into lakes, rivers, or aquifers. The water in lakes, rivers, and aquifers then either evaporates back to the atmosphere or eventually flows back to the ocean, completing a cycle.

## Branches of Hydrology

Chemical hydrology is the study of the chemical characteristics of water.

Ecohydrology is the study of interactions between organisms and the hydrologic cycle.

Hydrogeology is the study of the presence and movement of water in aquifers.

Hydroinformatics is the adaptation of information technology to hydrology and water resources applications.

Hydrometeorology is the study of the transfer of water and energy between land and water body surfaces and the lower atmosphere.

Isotope hydrology is the study of the isotopic signatures of water.

Surface hydrology is the study of hydrologic processes that operate at or near the Earth's surface.

## Related Fields

- Aquatic chemistry
- Civil engineering
- Climatology
- Environmental engineering
- Geomorphology
- Hydrography
- Hydraulic engineering
- Limnology
- Oceanography
- Physical Geography

## Hydrologic Measurements

The movement of water through the Earth can be measured in a number of ways. This information is important for both assessing water resources and understanding the processes involved in the hydrologic cycle. Following is a list of devices used by hydrologists and what they are used to measure.

- Disdrometer-precipitation characteristics
- Evaporation-Symon's evaporation pan
- Infiltrometer-infiltration
- Piezometer-groundwater pressure and, by inference, groundwater depth (see: aquifer test)
- Radar-cloud properties
- Rain gauge-rain and snowfall

- Satellite
- Sling psychrometer-humidity
- Stream gauge-stream flow (see: discharge (hydrology))
- Tensiometer-soil moisture
- Time domain reflectometer-soil moisture

## Hydrologic Prediction

Observations of hydrologic processes are used to make predictions of the future behaviour of hydrologic systems (water flow, water quality). One of the major current concerns in hydrologic research is the Prediction in Ungauged Basins (PUB), i.e. in basins where no or only very few data exist.

## Statistical Hydrology

By analysing the statistical properties of hydrologic records, such as rainfall or river flow, hydrologists can estimate future hydrologic phenomena. This, however, assumes the characteristics of the processes remain unchanged.

These estimates are important for engineers and economists so that proper risk analysis can be performed to influence investment decisions in future infrastructure and to determine the yield reliability characteristics of water supply systems. Statistical information is utilised to formulate operating rules for large dams forming part of systems which include agricultural, industrial and residential demands.

## Hydrologic Modeling

Hydrologic models are simplified, conceptual representations of a part of the hydrologic cycle. They are primarily used for hydrologic prediction and for understanding hydrologic processes. Two major types of hydrologic models can be distinguished:

- Models based on data. These models are black box systems, using mathematical and statistical concepts to link a certain input (for instance rainfall) to the model output (for instance runoff). Commonly used techniques are regression, transfer functions, neural networks and system identification. These models are known as stochastic hydrology models.

- Models based on process descriptions. These models try to represent the physical processes observed in the real world. Typically, such models contain representations of surface runoff, subsurface flow, evapotranspiration, and channel flow, but they can be far more complicated. These models are known as deterministic hydrology models. Deterministic hydrology models can be subdivided into single-event models and continuous simulation models.

Recent research in hydrologic modeling tries to have a more global approach to the understanding of the behaviour of hydrologic systems to make better predictions and to face the major challenges in water resources management.

**Hydrologic Transport**

Water movement is a significant means by which other material, such as soil or pollutants, are transported from place to place. Initial input to receiving waters may arise from a point source discharge or a line source or area source, such as surface runoff. Since the 1960s rather complex mathematical models have been developed, facilitated by the availability of high speed computers. The most common pollutant classes analyzed are nutrients, pesticides, total dissolved solids and sediment.

**Applications of Hydrology**

- Determining the water balance of a region
- Designing riparian restoration projects.
- Mitigating and predicting flood, landslide and drought risk;
- Designing irrigation schemes and managing agricultural productivity;
- Part of the hazard module in catastrophe modeling
- Providing drinking water;
- Designing dams for water supply or hydroelectric power generation;

- Designing bridges;
- Designing sewers and urban drainage system
- Analyzing the impacts of antecedent moisture on sanitary sewer systems.
- Predicting geomorphological changes, such as erosion or sedimentation.
- Assessing the impacts of natural and anthropogenic environmental change on water resources.
- Assessing contaminant transport risk and establishing environmental policy guidelines.

## HYDROGRAPHY

Hydrography focuses on the measurement of physical characteristics of waters and marginal land. In the generalized usage, "hydrography" pertains to measurement and description of any waters. With that usage oceanography and limnology are subsets of hydrography. In specialized usage the term applies to those measurements and descriptions of navigable waters necessary for safe navigation of vessels.

### Overview

Large scale hydrography is usually undertaken by national or international organizations that sponsor data collection through precise surveys and the publication of charts and descriptive material for navigational purposes. The science of oceanography is, in part, an outgrowth of classical hydrography. In many respects the data are interchangeable, but marine hydrographic data will be particularly directed toward marine navigation and safety of that navigation. Marine resource exploration and exploitation is a significant application of hydrography, principally focussed on the search for hydrocarbons.

Hydrographic measurements will include the tidal, current and wave information of physical oceanography. They will include bottom measurements, with particular emphasis on those marine geographical features that pose a hazard to navigation such as rocks, shoals, reefs and other features that obstruct ship passage. Unlike oceanography, hydrography will include shore features, natural and manmade, that aid in navigation. A hydrographic survey will therefore include accurate positions and representations of hills, mountains and even lights and towers that will aid in fixing a ship's position as well as the aspects of the sea and seabed.

Hydrography, partly for reasons of safety, tends to be more traditional in outlook and has conventions that are not entirely "scientific" in some views.

For example, hydrographic charts will usually tend to over represent least depths and ignore the actual submarine topography that will be portrayed on bathymetric charts. The former are the mariner's tools to avoid accident. The latter are best representations of the actual seabed, as in a topographic map, for scientific and other purposes.

A hydrographic survey differs from a bathymetric survey in some important respects, particularly in a bias toward least depths, because of the safety requirements of the former and geomorphologic descriptive requirements of the latter. As just one important example the echosoundings will be conducted under settings biased toward least depths while in bathymetric surveys they will be set for best description of the submarine topographical features that may include sound velocity and slope corrections that are more accurate but eliminate the safety bias.

Hydrography of streams will include information on the stream bed, flows, water quality and surrounding land. Basin or Interior Hydrography pays special attention to rivers and potable water.

## History

Hydrography's origin lies in the making of maps done by geographers in by means drawings and notations made by individual mariners. These were usually the private property, even closely held secrets, of individuals who used them for commercial or military advantage. Eventually organizations, particularly navies, realized the collection of this individualized knowledge and distribution to their members gave an organizational advantage. The next step was to organize members to actively collect information. Thus were born dedicated hydrographic organizations for the collection, organization, publication and distribution of hydrography incorporated into charts and sailing directions.

An interesting historical relationship is that of James Whistler to hydrography. His artistic talents were applied to the sometimes beautiful shore profiles that appeared on charts during his work as a cartographer with both the civilian and naval U. S. hydrographic organizations. Those profiles on early charts were etchings designed to aid mariners in identifying their landfall and harbor approaches.

## Organisations

Hydrographic services in most countries are carried out by specialised hydrographic offices. The international coordination of hydrographic efforts lies with the International Hydrographic Organization.

## LIMNOLOGY

Limnology is the study of inland waters (fresh), including their biological, physical, chemical, geological and hydrological aspects. This includes the study of (natural and man-made) lakes and ponds, rivers and streams, wetlands and groundwaters.

François-Alphonse Forel (1841-1912) established the field with his studies of Lake Geneva. Limnology traditionally is closely related to hydrobiology, which is concerned with the application of the principles and methods of physics, chemistry, geology, and geography to ecological problems.

## Landscape Ecology

Landscape ecology is a sub-discipline of ecology and geography that address how spatial variation in the landscape affects ecological processes such as the distribution and flow of energy, materials and individuals in the environment (which, in turn, may influence the distribution of landscape "elements" themselves such as hedgerows). Landscape ecology typically deals with problems in an applied and holistic context.

## Terminology

The term landscape ecology was coined by Carl Troll, a German geographer in 1939 (Troll 1939). He developed this terminology and many early concepts of landscape ecology as part of his early work applying aerial photograph interpretation to studies of interactions between environment and vegetation.

## Explanation

Heterogeneity is the measure of how different parts of a landscape are from one another. Landscape ecology looks at how spatial structure affects organism abundance at the landscape level, as well as the behavior and functioning of the landscape as a whole. This includes the study of the pattern, or the internal order of a landscape, on process, or the continuous operation of functions of organisms (Turner 1989). Landscape ecology also includes geomorphology as applied to the design and architecture of landscapes (Allaby 1998). Geomorphology is the study of how geological formations are responsible for the structure of a landscape.

## Evolution of Theory

One central landscape ecology theory originated from MacArthur & Wilson's The Theory of Island Biogeography. This work considered the assembly of flora and fauna on islands as the result of colonization from a mainland stock and stochastic extinction. The concepts of island biogeography were generalized from physical islands to abstract patches of habitat by Levins' metapopulation model. This generalization spurred the growth of landscape ecology by providing conservation biologists a new tool to assess how habitat fragmentation affects population viability. Recent growth of landscape ecology owes much to the development of geographic information systems (GIS) technology and the availability of large-extent habitat data (e.g. remotely sensed satellite images or aerial photography).

## Development as a Discipline

Landscape ecology developed in Europe from historical planning on human-dominated landscapes. In North America, concepts from general ecology theory were integrated. While general ecology theory and its sub-disciplines focused on the study of more homogenous, discrete community units organized in a hierarchical structure (typically as populations, species, and communities), landscape ecology built upon heterogeneity in space and time, and frequently included human-caused landscape changes in theory and application of concepts (Sanderson and Harris 2000).

By 1980, landscape ecology was a discrete, established discipline, marked by the organization of the International Association for Landscape Ecology (IALE) in 1982 and landmark book publications defining the scope and goals of the discipline, including Naveh and Lieberman (1984) and Forman and Godron (1986) (Ryszkowski 2002). Forman (1995) wrote that although study of "the ecology of spatial configuration at the human scale" was barely a decade old, there was strong potential for theory development and application of the conceptual framework. Today, theory and application of landscape ecology continues to develop through a need for innovative applications in a changing landscape and environment. Landscape ecology today relies more on advanced technologies such as remote sensing, GIS, and simulation models, with associated development of powerful quantitative methods to examine the interactions of patterns and processes (Turner et al. 2001). An example would be determining the amount of carbon present in the soil based on landform over a landscape, derived from GIS maps, vegetation types, and rainfall data for a region.

## Relationship to Ecological Theory

Although landscape ecology theory may be slightly outside of the "classical and preferred domain of scientific disciplines" because of the large, heterogeneous areas of study (Sanderson and Harris 2000), general ecology theory is central to landscape ecology theory in many aspects. Landscape ecology is comprised of four main principles, which include:

1. the development and dynamics of spatial heterogeneity,
2. interactions and exchanges across heterogeneous landscapes,
3. influences of spatial heterogeneity on biotic and abiotic processes, and

4. the management of spatial heterogeneity. The main difference from traditional ecological studies, which frequently assume that systems are spatially homogenous, is the consideration of spatial patterns (Turner and Gardner 1991).

## Important Terms in Landscape Ecology

Landscape ecology not only embraced a new vocabulary of terms but also incorporated general ecology theory terms in new ways. Many of the terms used in landscape ecology are as interconnected and interrelated as the discipline itself. Landscape can be defined as an area containing two or more ecosystems in close proximity (Sanderson and Harris 2000).

## Scale and Heterogeneity (incorporating composition, structure, and function)

A main concept in landscape ecology is scale. Scale represents the real world as translated onto a map, in the relationship between distance on a map image and the corresponding distance on earth (Malczewski 1999). Scale is also the spatial or temporal measure of an object or a process (Turner and Gardner 1991), or level or degree of spatial resolution (Forman 1995). Components of scale include composition, structure, and function, which are all important ecological concepts. Applied to landscape ecology, composition refers to the number of patch types (see below) represented on a landscape, and their relative abundance. For example, the amount of forest or wetland, the length of forest edge, or the density of roads can be aspects of landscape composition. Structure is determined by the composition, the configuration, and the proportion of different patches across the landscape, while function refers to how each element in the landscape interacts based on its life cycle events (Turner and Gardner 1991). Pattern is the term for the contents and internal order of a heterogeneous area of land (Forman and Godron 1986).

A landscape with structure and pattern implies that it has spatial heterogeneity, or the uneven, non-random distribution of objects across the landscape (Forman 1995). Heterogeneity is a key element of landscape ecology that separates this discipline from other branches of ecology.

## Patch and Mosaic

Patch, a term fundamental to landscape ecology, is defined as a relatively homogeneous area that differs from its surroundings (Forman 1995). Patches are the basic unit of the landscape that change and fluctuate, a process called patch dynamics. Patches have a definite shape and spatial configuration, and can be described compositionally by internal variables such as number of trees, number of tree species, height of trees, or other similar measurements (Forman 1995).

Matrix is the "background ecological system" of a landscape with a high degree of connectivity. Connectivity is the measure of how connected or spatially continuous a corridor, network, or matrix is (Forman 1995). For example, a forested landscape (the matrix) with fewer gaps in forest cover (open patches) will have higher connectivity. Corridors have important functions as strips of a particular type of landscape differing from adjacent land on both sides (Forman 1995). A network is an interconnected system of corridors while mosaic describes the pattern of patches, corridors and matrix that form a landscape in its entirety (Forman 1995).

## Boundary and Edge

Landscape patches have a boundary between them which can be defined or fuzzy (Sanderson and Harris 2000). The zone composed of the edges of adjacent ecosystems is the boundary (Forman 1995). Edge means the portion of an ecosystem near its perimeter, where influences of the adjacent patches can cause an environmental difference between the interior of the patch and its edge.

This edge effect includes a distinctive species composition or abundance in the outer part of the landscape patch (Forman 1995). For example, when a landscape is a mosaic of perceptibly different types, such as a forest adjacent to a grassland, the edge is the location where the two types adjoin. In a continuous landscape, such as a forest giving way to open woodland, the exact edge location is fuzzy and is sometimes determined by a local gradient exceeding a threshold, such as the point where the tree cover falls below thirty-five percent (Turner and Gardner 1991).

### Ecotones, Ecoclines, and Ecotopes

A type of boundary is the ecotone, or the transitional zone between two communities (Allaby 1998). Ecotones can arise naturally, such as a lakeshore, or can be human-created, such as a cleared agricultural field from a forest (Allaby 1998). The ecotonal community retains characteristics of each bordering community and often contains species not found in the adjacent communities. Classic examples of ecotones include fencerows; forest to marshlands transitions; forest to grassland transitions; or land-water interfaces such as riparian zones in forests. Characteristics of ecotones include vegetational sharpness, physiognomic change, occurrence of a spatial community mosaic, many exotic species, ecotonal species, spatial mass effect, and species richness higher or lower than either side of the ecotone (Walker et al. 2003). An ecocline is another type of landscape boundary, but it is a gradual and continuous change in envircnmental conditions of an ecosystem or community. Ecoclines help explain the distribution and diversity of organisms within a landscape because certain organisms survive better under certain conditions, which change along the ecocline. They contain heterogeneous communities which are considered more environmentally stable than those of ecotones (Attrill and Rundle 2002).

An ecotope is a spatial term representing the smallest ecologically-distinct unit in mapping and classification of landscapes (Forman 1995). Relatively homogeneous, they are spatially-explicit landscape units used to stratify landscapes into ecologically distinct features for measurement and mapping of landscape structure, function, and change over time, and to examine the effects of disturbance and fragmentation.

### Disturbance and Fragmentation

Disturbance is an event that significantly alters the pattern of variation in the structure or function of a system, while fragmentation is the breaking up of a habitat, ecosystem, or land-use type into smaller parcels (Forman 1995). Disturbance is generally considered a natural process. Fragmentation causes land transformation, an important current process in landscapes as more and more development occurs.

## Landscape Ecology Theory

### *Elements of Landscape Ecology Theory*

Landscape ecology, as a theory, stresses the role of human impacts on landscape structures and functions and proposes ways for restoring degraded landscapes (Naveh and Lieberman 1984). Landscape ecology explicitly includes humans as entities that cause functional changes on the landscape (Sanderson and Harris 2000). Landscape ecology theory includes the landscape stability principle, which emphasizes the importance of landscape structural heterogeneity in developing resistance to disturbances, recovery from disturbances, and promoting total system stability (Forman and Godron 1986).

This principle is a major contribution to general ecological theories which highlight the importance of relationships among the various components of the landscape. Integrity of landscape components helps maintain resistance to external threats, including development and land transformation by human activity (Turner et al. 2001). Analysis of land use changes has included a strongly geographical approach within landscape ecology. This has lead to acceptance of the idea of multifunctional properties of landscapes (Ryszkowski 2002). There are still calls for a more unified theory of landscape ecology due to differences in professional opinion among landscape ecologists, and the interdisciplinary approach to the discipline (Bastian 2001).

An important related theory is hierarchy theory, which refers to how systems of discrete functional elements operate when linked at two or more scales. For example, a forested landscape might be hierarchically composed of drainage basins, which in turn are composed of local ecosystems or stands, which are in turn composed of individual trees and tree gaps (Forman 1995). Recent theoretical developments in landscape ecology have emphasized the relationship between pattern and process, as well as the effect that changes in spatial scale has on the potential to extrapolate information across scales (Turner and Gardner 1991). Several studies suggest that the landscape has critical thresholds at which ecological processes will show dramatic changes, such as the complete transformation of a landscape by an invasive species with a small change in average temperatures per year which favors the invasive habitat requirements (Turner and Gardner 1991).

## Landscape Ecology Application

### *Research Directions*

Developments in landscape ecology illustrate the important relationships between spatial patterns and ecological processes, and incorporate quantitative methods that link spatial patterns and ecological processes at broad spatial and temporal scales. This linkage of time, space, and environmental change can assist land managers in applying land management plans to solve environmental problems (Turner et al. 2001). The increased attention in recent years on spatial dynamics has highlighted the need for new quantitative methods that can analyze patterns, determine the importance of spatially explicit processes on the landscape, and develop reliable landscape models (Turner and Gardner 1991). Multivariate analysis techniques, a type of statistics incorporating many variables, are frequently used to examine landscape level vegetation patterns. A number of studies in riparian systems and wetlands use a variety of statistical techniques, such as cluster analysis, canonical correspondence analysis (CCA), or detrended correspondence analysis (DCA), for classifying vegetation. Gradient analysis is another way to determine the vegetation structure across a landscape, or to help delineate critical wetland habitat for conservation or mitigation purposes (Lyon and Sagers 1998, Choesin and Boerner 2002).

Climate change is another major component in structuring current research in landscape ecology. Ecotones, as a basic unit in landscape studies, may have significance for management under climate change scenarios, since change effects are likely to be seen at ecotones first because of the unstable nature of a fringe habitat (Walker et al. 2003). Research in northern regions has examined landscape ecological processes, such as the accumulation of snow during winter, snow melting, freeze-thaw action, percolation, soil moisture variation, and temperature regimes through long-term measurements in Norway (Loffler and Finch 2005). The study analyzes gradients across space and time between ecosystems of the central high mountains to determine relationships between distribution patterns of animals in their environment. Looking at where animals live, and how vegetation shifts over time, may provide insight into changes in snow and ice over long periods of time across the landscape as a whole.

Other landscape-scale studies maintain that human impact is likely the main determinant of landscape pattern over much of the globe (Wilson and King 1995). Landscapes may become substitutes for biodiversity measures because plant and animal composition differs between samples taken from sites within different landscape categories. Taxa, or different species, can "leak" from one habitat into another, which has implications for landscape ecology. As human land use practices expand, and continue to increase the proportion of edges in landscapes, the effects of this leakage across edges on assemblage integrity may become more significant and important in conservation because taxa may be conserved across landscape levels, if not at local levels (Dangerfield et al. 2003).

## Relationship to Other Disciplines

Landscape ecology has important links to application-oriented disciplines such as agriculture and forestry. In agriculture, landscape ecology has introduced new options for the control and management of environmental threats brought about by the intensification of agricultural practices. Agriculture has always been a strong human impact on ecosystems (Ryszkowski 2002). In forestry, changes in consumer needs have changed conservation and use of forested landscapes from structuring stands for fuelwood and timber to ordering stands across landscapes to enhance aesthetics, habitats and biological diversity. Landscape forestry provides methods, concepts, and analytic procedures for shifting management from traditional to landscape forestry (Boyce 1995). Landscape ecology has been cited as a major contributor to the development of fisheries biology as a distinct biological science discipline (Magnuson 1991), and is frequently incorporated in study design for wetland delineation in hydrology (Attrill and Rundle 2002).

# 4

# OCEANOGRAPHY

## INTRODUCTION

Oceanography (from Ocean + Greek fe = write), also called oceanology or marine science, is the branch of Earth Sciences that studies the Earth's oceans and seas. It covers a wide range of topics, including marine organisms and ecosystem dynamics; ocean currents, waves, and geophysical fluid dynamics; plate tectonics and the geology of the sea floor; and fluxes of various chemical substances and physical properties within the ocean and across its boundaries.

These diverse topics reflect multiple disciplines that oceanographers blend to further knowledge of the world ocean and understanding of processes within it: biology, chemistry, geology, meteorology, and physics.

**Branches:** The study of oceanography may be divided into a number of branches:

- Marine biology or biological oceanography, the study of the plants, animals and microbes (biota) of the oceans and their ecological interaction;
- Chemical oceanography or marine chemistry, the study of the chemistry of the ocean and its chemical interaction with the atmosphere;
- Marine geology or geological oceanography, the study of the geology of the ocean floor including plate tectonics;
- Physical oceanography studies the ocean's physical attributes including temperature-salinity structure, mixing, waves, tides and currents;

- Marine engineering involves the design and building of oil platforms, ships, harbors, and other structures that allow us to use the ocean wisely.

These branches reflect the fact that many oceanographers are first trained in the exact sciences or mathematics and then focus on applying their interdisciplinary knowledge, skills and abilities to oceanography.'

## HISTORY

Early exploration of the oceans was limited to its surfaces and the few creatures that fishermen brought up in nets, but when Louis Antoine de Bougainville and James Cook carried out their explorations in the South Pacific, the seas themselves formed part of the reports.

James Rennell wrote the first scientific textbooks about currents in the Atlantic and Indian oceans during the late 18th and at the beginning of 19th century. Sir James Clark Ross took the first modern sounding in deep sea in 1840, and Charles Darwin published a paper on reefs and the formation of atolls.

The steep slope beyond the continental shelves was not discovered until 1849. Matthew Fontaine Maury's Physical Geography of the Sea, 1855 was the first textbook of oceanography. The first successful laying of Transatlantic telegraph cable in August 1858 confirmed the presence of an underwater "telegraphic plateau" mid-ocean ridge.

After the middle of the 19th century, scientific societies were processing a flood of new terrestrial botanical and zoological information. European natural historians began to sense the lack of more than anecdotal knowledge of the oceans.

In 1871, Under the recommendations of the Royal Society of London, the British government sponsored an expedition to explore world's oceans and conduct scientific investigations. With that, oceanography began as a quantifiable science in 1872, when the Scots Charles Wyville Thompson and Sir John Murray launched the Challenger expedition (1872-1876). Other European and American nations also sent out scientific expeditions (as did private individuals and institutions). The four-month 1910 North Atlantic expedition headed by Sir John

Murray and Johan Hjort was at that time the most ambitious research oceanographic and marine zoological project ever, and led to the classic 1912 book The Depths of the Ocean.

Oceanographic institutes dedicated to the study of oceanography were founded. In the United States, these included the Scripps Institution of Oceanography, Woods Hole Oceanographic Institution, Lamont-Doherty Earth Observatory at Columbia University, and the School of Oceanography at University of Washington. In Britain, there is a major research institution: National Oceanography Centre, Southampton. In Australia, CSIRO Marine and Atmospheric Research, known as CMAR, is a leading center.

The first international organization of oceanography was created in 1902 as the International Council for the Exploration of the Sea.

In 1921 Monaco formed the International Hydrographic Bureau (IHB). Then in 1966, the U.S. Congress created a National Council for Marine Resources and Engineering Development. NOAA was in charge of exploring and studying all aspects of Oceanography. It also enabled the National Science Foundation to award Sea Grant College funding to multi-disciplinary researchers in the field of oceanography.

### Ocean and Atmosphere Connections

The study of the oceans is intimately linked to understanding global warming and related biosphere concerns.

> *"Our planet is invested with two great oceans; one visible, the other invisible; one underfoot, the other overhead; one entirely envelopes it, the other covers about two thirds of its surface." - Matthew F. Maury (1855) The Physical Geography of the Seas and Its Meteorology*

## PEDOLOGY (SOIL STUDY)

Pedology is the study of soils in its natural environment. It is one of two main branches of soil science, the other being edaphology. Pedology deals with pedogenesis, soil morphology, soil classification.

## Overview

Soil is not only a support for vegetation, but it is also the zone (the pedosphere) of numerous interactions between climate (water, air, temperature), soil life (micro-organisms, plants, animals) and its residues, the mineral material of the original and added rock, and its position in the landscape. During its formation and genesis, the soil profile slowly deepens and develops characteristic layers, called 'horizons', while a steady state balance is approached.

Soil users (such as agronomists) showed initially little concern in the dynamics of soil. They saw it as medium whose chemical, physical and biological properties were useful for the services of agronomic productivity. On the other hand, pedologists and geologists did not initially focus on the agronomic applications of the soil characteristics (edaphic properties) but upon its relation to the nature and history of landscapes. Today, there's an integration of the two disciplinary approaches as part of landscape and environmental sciences.

Pedologists are now also interested in the practical applications of a good understanding of pedogenesis processes (the evolution and functioning of soils), like interpreting its environmental history and predicting consequences of changes in land use, while agronomists understand that the cultivated soil is a complex medium, often resulting from several thousands of years of evolution.

They understand that the current balance is fragile and that only a thorough knowledge of its history makes it possible to ensure its sustainable use.

## Concepts

- Complexity in soil genesis is more common than simplicity.
- Soils lie at the interface of Earth's atmosphere, biosphere, hydrosphere and lithosphere. Therefore, a thorough understanding of soils requires some knowledge of meteorology, climatology, ecology, biology, hydrology, geomorphology, geology and many other earth sciences and natural sciences.

- Contemporary soils carry imprints of pedogenic processes that were active in the past, although in many cases these imprints are difficult to observe or quantify. Thus, knowledge of paleoecology, paleogeography, glacial geology and paleoclimatology is important for the recognition and understanding of soil genesis and constitute a basis for predicting the future soil changes.
- Five major, external factors of soil formation (climate, organisms, relief, parent material and time), and several smaller, less identifiable ones, drive pedogenic processes and create soil patterns.
- Characteristics of soils and soil landscapes, e.g., the number, sizes, shapes and arrangements of soil bodies, each of which is characterized on the basis of soil horizons, degree of internal homogeneity, slope, aspect, landscape position, age and other properties and relationships, can be observed and measured.
- Distinctive bioclimatic regimes or combinations of pedogenic processes produce distinctive soils. Thus, distinctive, observable morphological features, e.g., illuvial clay accumulation in B horizons, are produced by certain combinations of pedogenic processes operative over varying periods of time.
- Pedogenic (soil-forming) processes act to both create and destroy order (anisotropy) within soils; these processes can proceed simultaneously. The resulting soil profile reflects the balance of these processes, present and past.
- The geological Principle of Uniformitarianism applies to soils, i.e., pedogenic processes active in soils today have been operating for long periods of time, back to the time of appearance of organisms on the land surface. These processes do, however, have varying degrees of expression and intensity over space and time.
- A succession of different soils may have developed, eroded and/or regressed at any particular site, as soil genetic factors and site factors, e.g., vegetation, sedimentation, geomorphology, change.

- There are very few old soils (in a geological sense) because they can be destroyed or buried by geological events, or modified by shifts in climate by virtue of their vulnerable position at the surface of the earth. Little of the soil continuum dates back beyond the Tertiary period and most soils and land surfaces are no older than the Pleistocene Epoch. However, preserved/lithified soils (paleosols) are an almost ubiqutious feature in terrestrial (land-based) environments throughout most of geologic time. Since they record evidence of ancient climate change, they present immense utility in understanding climate evolution throughout geologic history.
- Knowledge and understanding of the genesis of a soil is important in its classification and mapping.
- Soil classification systems cannot be based entirely on perceptions of genesis, however, because genetic processes are seldom observed and because pedogenic processes change over time.
- Knowledge of soil genesis is imperative and basic to soil use and management. Human influence on, or adjustment to, the factors and processes of soil formation can be best controlled and planned using knowledge about soil genesis.
- Soils are natural clay factories (clay includes both clay mineral structures and particles less than 2 μm in diameter). Shales worldwide are, to a considerable extent, simply soil clays that have been formed in the pedosphere and eroded and deposited in the ocean basins, to become lithified at a later date.

## PALEOGEOGRAPHY

Paleogeography (sometimes spelled paleogeography) is the study of the ancient geologic environments of the Earth's surface as preserved in the stratigraphic record.

Paleogeographic analysis is used in the detailed study of sedimentary basins in petroleum geology. Paleogeographers also study the sedimentary environment associated with fossils to aid in the understanding of evolutionary development of extinct species. The reconstructions of prehistoric continents and oceans depends on paleogeographic evidence. Thus paleogeography provided critical evidence for the development of continental drift and current plate tectonic theories. For example, knowledge of the shape and latitudinal location of supercontinents such as Pangaea and ancient oceans such as Panthalassa result from paleogeographic studies.

## Plate Tectonics

Plate tectonics (from Greek, tekton "builder" or "mason") is a theory of geology that has been developed to explain the observed evidence for large scale motions of the Earth's lithosphere. The theory encompassed and superseded the older theory of continental drift from the first half of the 20th century and the concept of seafloor spreading developed during the 1960s.

The outermost part of the Earth's interior is made up of two layers: above is the lithosphere, comprising the crust and the rigid uppermost part of the mantle. Below the lithosphere lies the asthenosphere. Although solid, the asthenosphere has relatively low viscosity and shear strength and can flow like a liquid on geological time scales. The deeper mantle below the asthenosphere is more rigid again. This is, however, not due to cooler temperatures but due to high pressure.

The lithosphere is broken up into what are called tectonic plates-in the case of Earth, there are seven major and many minor plates (see list below). The lithospheric plates ride on the asthenosphere. These plates move in relation to one another at one of three types of plate boundaries: convergent or collision boundaries, divergent or spreading boundaries, and transform boundaries. Earthquakes, volcanic activity, mountain-building, and oceanic trench formation occur along plate boundaries. The lateral movement of the plates is typically at speeds of 0.66 to 8.50 centimeters per year.

## Synopsis of the Development of the Theory

Plate tectonic theory arose out of the hypothesis of continental drift first proposed by Alfred Wegener in 1912 and expanded in his 1915 book The Origin of Continents and Oceans that said in part that all the continents once been together in a single land mass that had drifted apart. But it was not until the mechanic of seafloor spreading (first articulated by Robert S. Dietz, but Harry Hess is usually given credit (see below)) in the early 1960s that the theory actually became accepted by the scientific community.

Following the recognition of magnetic anomalies defined by symmetric, parallel stripes of similar magnetization on the seafloor on either side of a mid-ocean ridge, plate tectonics quickly became broadly accepted. Simultaneous advances in early seismic imaging techniques in and around Wadati-Benioff zones collectively with numerous other geologic observations soon solidified plate tectonics as a theory with extraordinary explanatory and predictive power.

Study of the deep ocean floor was critical to development of the theory; the field of deep sea marine geology accelerated in the 1960s. Correspondingly, plate tectonic theory was developed during the late 1960s and has since been essentially universally accepted by scientists throughout all geoscientific disciplines. The theory has revolutionized the Earth sciences because of its unifying and explanatory power for diverse geological phenomena.

## Key Principles

The division of the outer parts of the Earth's interior into lithosphere and asthenosphere is based on their mechanical differences and in the ways that heat is transferred. The lithosphere is cooler and more rigid, whilst the asthenosphere is hotter and mechanically weaker. Also, the lithosphere loses heat by conduction whereas asthenosphere transfers heat by convection and has a nearly adiabatic temperature gradient. This division should not be confused with the chemical subdivision of the Earth into (from innermost to outermost) core, mantle, and crust. The lithosphere contains both crust and some mantle. A given piece of mantle may be part of the lithosphere or the asthenosphere at different times, depending on its temperature, pressure and shear strength. The key principle of plate tectonics is that the lithosphere exists as separate and distinct tectonic plates, which ride on the fluid-like (visco-elastic solid) asthenosphere. Plate motions range from a few millimeters per year (about as fast as our fingernails grow) to about 15 centimeters per year (about as fast as our hair grows). The plates are around 100 km (60 miles) thick and consist of lithospheric mantle overlain by either of two types of crustal material: oceanic crust (in older texts called sima from silicon and magnesium) and continental crust (sial from silicon and aluminium). The two types of crust differ in thickness, with continental crust considerably thicker than oceanic (50 km vs 5 km).

One plate meets another along a plate boundary, and plate boundaries are commonly associated with geological events such as earthquakes and the creation of topographic features like mountains, volcanoes and oceanic trenches. The majority of the world's active volcanoes occur along plate boundaries, with the Pacific Plate's Ring of Fire being most active and famous. These boundaries are discussed in further detail below.

Tectonic plates can include continental crust or oceanic crust, and typically, a single plate carries both. For example, the African Plate includes the continent and parts of the floor of the Atlantic and Indian Oceans. The distinction between continental crust and oceanic crust is based on the density of constituent materials; oceanic crust is denser than continental crust owing to their different proportions of various elements, particularly, silicon. Oceanic crust is denser because it has less silicon and more heavier elements ("mafic") than continental crust ("felsic"). As a result, oceanic crust generally lies below sea level (for example most of the Pacific Plate), while the continental crust projects above sea level (see isostasy for explanation of this principle).

## Types of Plate Boundaries

Three types of plate boundaries exist, characterized by the way the plates mo e relative to each other. They are associated with different types of surface phenomena. The different types of plate boundaries are:

1. Transform boundaries occur where plates slide or, perhaps more accurately, grind past each other along transform faults. The relative motion of the two plates

is either sinistral (left side toward the observer) or dextral (right side toward the observer). The San Andreas Fault in California is one example.

2. Divergent boundaries occur where two plates slide apart from each other. Mid-ocean ridges (e.g., Mid-Atlantic Ridge) and active zones of rifting (such as Africa's Great Rift Valley) are both examples of divergent boundaries.
3. Convergent boundaries (or active margins) occur where two plates slide towards each other commonly forming either a subduction zone (if one plate moves underneath the other) or a continental collision (if the two plates contain continental crust). Deep marine trenches are typically associated with subduction zones. Because of friction and heating of the subducting slab, volcanism is almost always closely linked. Examples of this are the Andes mountain range in South America and the Japanese island arc.

**Transform (conservative) Boundaries**

John Tuzo Wilson recognized that because of friction, the plates cannot simply glide past each other. Rather, stress builds up in both plates and when it reaches a level that exceeds the strain threshold of rocks on either side of the fault the accumulated potential energy is released as strain. Strain is both accumulative and/or instantaneous depending on the rheology of the rock; the ductile lower crust and mantle accumulates deformation gradually via shearing whereas the brittle upper crust reacts by fracture, or instantaneous stress release to cause motion along the fault. The ductile surface of the fault can also release instantaneously when the strain rate is too great. The energy released by instantaneous strain release is the cause of earthquakes, a common phenomenon along transform boundaries.

A good example of this type of plate boundary is the San Andreas Fault which is found in the western coast of North America and is one part of a highly complex system of faults in this area. At this location, the Pacific and North American plates move relative to each other such that the Pacific plate is moving northwest with respect to North America. Other examples of transform faults include the Alpine Fault in New Zealand and the North Anatolian Fault in Turkey. Transform faults are also found offsetting the crests of mid-ocean ridges (for example, the Mendocino Fracture Zone offshore northern California).

### Divergent (constructive) Boundaries

Bridge across the Álfagjá rift valley near Grindavik on the Reykjanes peninsula in southwest Iceland, the boundary of the Eurasian and North American continental tectonic plates.

At divergent boundaries, two plates move apart from each other and the space that this creates is filled with new crustal material sourced from molten magma that forms below. The origin of new divergent boundaries at triple junctions is sometimes thought to be associated with the phenomenon known as hotspots. Here, exceedingly large convective cells bring very large quantities of hot asthenospheric material near the surface and the kinetic energy is thought to be sufficient to break apart the lithosphere. The hot spot which may have initiated the Mid-Atlantic Ridge system currently underlies Iceland which is widening at a rate of a few centimeters per year.

Divergent boundaries are typified in the oceanic lithosphere by the rifts of the oceanic ridge system, including the Mid-Atlantic Ridge and the East Pacific Rise, and in the continental lithosphere by rift valleys such as the famous East African Great Rift Valley. Divergent boundaries can create massive fault zones in the oceanic ridge system. Spreading is generally not uniform, so where spreading rates of adjacent ridge blocks are different, massive transform faults occur. These are the fracture zones, many bearing names, that are a major source of submarine earthquakes.

A sea floor map will show a rather strange pattern of blocky structures that are separated by linear features perpendicular to the ridge axis. If one views the sea floor between the fracture zones as conveyor belts carrying the ridge on each side of the rift away from the spreading center the action becomes clear. Crest depths of the old ridges, parallel to the current spreading center, will be older and deeper (from thermal contraction and subsidence).

It is at mid-ocean ridges that one of the key pieces of evidence forcing acceptance of the sea-floor spreading hypothesis was found. Airborne geomagnetic surveys showed a strange pattern of symmetrical magnetic reversals on opposite sides of ridge centers. The pattern was far too regular to be coincidental as the widths of the opposing bands were too closely matched. Scientists had been studying polar reversals and the link was made by Lawrence W. Morley, Frederick John Vine and Drummond Hoyle Matthews in the Morley-Vine-Matthews Hypothesis. The magnetic banding directly corresponds with the Earth's polar reversals. This was confirmed by measuring the ages of the rocks within each band. The banding furnishes a map in time and space of both spreading rate and polar reversals.

## Convergent (destructive) Boundaries

The nature of a convergent boundary depends on the type of lithosphere in the plates that are colliding. Where a dense oceanic plate collides with a less-dense continental plate, the oceanic plate is typically thrust underneath because of the greater buoyancy of the continental lithosphere, forming a subduction zone. At the surface, the topographic expression is commonly an oceanic trench on the ocean side and a mountain range on the continental side. An example of a continental-oceanic subduction zone is the area along the western coast of South America where the oceanic Nazca Plate is being subducted beneath the continental South American Plate.

While the processes directly associated with the production of melts directly above downgoing plates producing surface volcanism is the subject of some debate in the geologic community, the general consensus from ongoing research suggests that the release of volatiles is the primary contributor. As the subducting plate descends, its temperature rises driving off volatiles (most importantly water) encased in the porous oceanic crust. As this water rises into the mantle of the overriding plate, it lowers the melting temperature of surrounding mantle, producing melts (magma) with large amounts of dissolved gases. These melts rise to the surface and are the source of some of the most explosive volcanism on Earth because of their high volumes of extremely pressurized gases (consider Mount St. Helens). The melts rise to the surface and cool forming long chains of volcanoes inland from the continental shelf and parallel to it.

The continental spine of western South America is dense with this type of volcanic mountain building from the subduction of the Nazca plate. In North America the Cascade mountain range, extending north from California's Sierra Nevada, is also of this type.

Such volcanoes are characterized by alternating periods of quiet and episodic eruptions that start with explosive gas expulsion with fine particles of glassy volcanic ash and spongy cinders, followed by a rebuilding phase with hot magma. The entire Pacific Ocean boundary is surrounded by long stretches of volcanoes and is known collectively as The Ring of Fire.

Where two continental plates collide the plates either buckle and compress or one plate delves under or (in some cases) overrides the other. Either action will create extensive mountain ranges. The most dramatic effect seen is where the northern margin of the Indian Plate is being thrust under a portion of the Eurasian plate, lifting it and creating the Himalayas and the Tibetan Plateau beyond. It has also caused parts of the Asian continent to deform westward and eastward on either side of the collision.

When two plates with oceanic crust converge they typically create an island arc as one plate is subducted below the other. The arc is formed from volcanoes which erupt through the overriding plate as the descending plate melts below it. The arc shape occurs because of the spherical surface of the earth (nick the peel of an orange with a knife and note the arc formed by the straight-edge of the knife). A deep undersea trench is located in front of such arcs where the descending slab dips downward. Good examples of this type of plate convergence would be Japan and the Aleutian Islands in Alaska.

Plates may collide at an oblique angle rather than head-on (e.g. one plate moving north, the other moving south-east), and this may cause strike-slip faulting along the collision zone, in addition to subduction.

Not all plate boundaries are easily defined. Some are broad belts whose movements are unclear to scientists. One example would be the Mediterranean-Alpine boundary, which involves two major plates and several micro plates. The boundaries of the plates do not necessarily coincide with those of the continents. For instance, the North American Plate covers not only North America, but also far eastern Siberia and northern Japan.

## Driving Forces of Plate Motion

Tectonic plates are able to move because of the relative density of oceanic lithosphere and the relative weakness of the asthenosphere. Dissipation of heat from the mantle is acknowledged to be the original source of energy driving plate tectonics, but it is no longer thought that the plates ride passively on asthenospheric convection currents. Instead, it is accepted that the excess density of the oceanic lithosphere sinking in subduction zones drives plate motions.

When it forms at mid-ocean ridges, the oceanic lithosphere is initially less dense than the underlying asthenosphere, but it becomes more dense with age, as it conductively cools and thickens.

The greater density of old lithosphere relative to the underlying asthenosphere allows it to sink into the deep mantle at subduction zones, providing most of the driving force for plate motions. The weakness of the asthenosphere allows the tectonic plates to move easily towards a subduction zone.

Two and three-dimensional imaging of the Earth's interior (seismic tomography) shows that there is a laterally heterogeneous density distribution throughout the mantle. Such density variations can be material (from rock chemistry), mineral (from variations in mineral structures), or thermal (through thermal expansion and contraction from heat energy). The manifestation of this lateral density heterogeneity is mantle convection from buoyancy forces.

How mantle convection relates directly and indirectly to the motion of the plates is a matter of ongoing study and discussion in geodynamics. Somehow, this energy must be transferred to the lithosphere in order for tectonic plates to move. There are essentially two types of forces that are thought to influence plate motion: friction and gravity.

## FRICTION

### Basal Drag

Large scale convection currents in the upper mantle are transmitted through the asthenosphere; motion is driven by friction between the asthenosphere and the lithosphere.

### Slab Suction

Local convection currents exert a downward frictional pull on plates in subduction zones at ocean trenches. Although, one could in effect argue that Slab-suction is actually merely a unique geodynamic setting wherein which basal tractions continue to act on the plate as it dives into the mantle (although perhaps to a greater extent-acting on both the under and upper side of the slab).

### Gravitation

Gravitational sliding: Plate motion is driven by the higher elevation of plates at ocean ridges. As oceanic lithosphere is formed at spreading ridges from hot mantle material it gradually cools and thickens with age (and thus distance from the ridge). Cool oceanic lithosphere is significantly denser than the hot mantle material from which it is derived and so with increasing thickness it gradually subsides into the mantle to compensate the greater load. The result is a slight lateral incline with distance from the ridge axis.

Casually in the geophysical community and more typically in the geological literature in lower education this process is often referred to as "ridge-push". This is, in fact, a misnomer as nothing is "pushing" and tensional features are dominant along ridges. It is more accurate to refer to this mechanism as gravitational sliding as variable topography across the totality of the plate can vary considerably and the topography of spreading ridges is only the most prominent feature. For example:

2. Flexural bulging of the lithosphere before it dives underneath an adjacent plate, for instance, produces a clear topographical feature that can offset or at least effect the influence of topographical ocean ridges.
3. Mantle plumes impinging on the underside of tectonic plates can drastically alter the topography of the ocean floor.

### Slab-pull

Plate motion is driven by the weight of cold, dense plates sinking into the mantle at trenches. There is considerable evidence that convection is occurring in the mantle at some scale. The upwelling of material at mid-ocean ridges is almost certainly part of this convection. Some early models of plate tectonics envisioned the plates riding on top of convection cells like conveyor belts.

However, most scientists working today believe that the asthenosphere is not strong enough to directly cause motion by the friction of such basal forces. Slab pull is most widely thought to be the greatest force acting on the plates. Recent models indicate that trench suction plays an important role as well. However, it should be noted that the North American Plate, for instance, is nowhere being subducted, yet it is in motion. Likewise the African, Eurasian and Antarctic Plates. The overall driving force for plate motion and its energy source remain subjects of on-going research.

### External Forces

In a study published in the January-February 2006 issue of the Geological Society of America Bulletin, a team of Italian and U.S. scientists argued that the westward component of plates is from Earth's rotation and consequent tidal friction of the moon. As the Earth spins eastward beneath the moon, they say, the moon's gravity ever so slightly pulls the Earth's surface layer back westward. It has also been suggested (albeit, controversially) that this observation may also explain why Venus and Mars have no plate tectonics since Venus has no moon, and Mars' moons are too small to have significant tidal effects on Mars. This is not, however, a new argument.

It was originally raised by the "father" of the plate tectonics hypothesis, Alfred Wegener. It was challenged by the physicist Harold Jeffreys who calculated that the magnitude of tidal friction required would have quickly brought the Earth's rotation to a halt long ago. Many plates are moving north and eastward, and the dominantly westward motion of the Pacific ocean basins is simply from the eastward bias of the Pacific spreading center (which is not a predicted manifestation of such lunar forces). It is argued, however, that relative to the lower mantle, there is a slight westward component in the motions of all the plates.

## Relative Significance of each Mechanism

The actual vector of a plate's motion must necessarily be a function of all the forces acting upon the plate. However, therein remains the problem of to what degree each process contributes to the motion of each tectonic plate. The diversity of geodynamic settings and properties of each plate must clearly result in differences in the degree to which such processes are actively driving the plates. One method of dealing with this problem is to consider the relative rate at which each plate is moving and to consider the available evidence of each driving force upon the plate as far as possible.

One of the most significant correlations found is that lithospheric plates attached to downgoing (subducting) plates move much faster than plates not attached to subducting plates. The Pacific plate, for instance, is essentially surrounded by zones of subduction (the so-called Ring of Fire) and moves much faster than the plates of the Atlantic basin, which are attached (perhaps one could say 'welded') to adjacent continents instead of subducting plates. It is thus thought that forces associated with the downgoing plate (slab pull and slab suction) are the driving forces which determine the motion of plates. The driving forces of plate motion are, nevertheless, still very active subjects of on-going discussion and research in the geophysical community.

## Major Plates

The main plates are:

- African Plate, covering Africa - Continental plate

- Antarctic Plate, covering Antarctica - Continental plate
- Australian Plate, covering Australia (fused with Indian Plate between 50 and 55 million years ago) - Continental plate
- Eurasian Plate covering Asia and Europe - Continental plate
- North American Plate covering North America and north-east Siberia - Continental plate
- South American Plate covering South America - Continental plate
- Pacific Plate, covering the Pacific Ocean - Oceanic plate

Notable minor plates include the Indian Plate, the Arabian Plate, the Caribbean Plate, the Juan de Fuca Plate, the Nazca Plate, the Philippine Plate and the Scotia Plate.

The movement of plates has caused the formation and break-up of continents over time, including occasional formation of a supercontinent that contains most or all of the continents. The supercontinent Rodinia is thought to have formed about 1 billion years ago and to have embodied most or all of Earth's continents, and broken up into eight continents around 600 million years ago. The eight continents later re-assembled into another supercontinent called Pangaea; Pangea eventually broke up into Laurasia (which became North America and Eurasia) and Gondwana (which became the remaining continents).

## HISTORICAL DEVELOPMENT OF THE THEORY

### Continental Drift

Continental drift was one of many ideas about tectonics proposed in the late 19th and early 20th centuries. The theory has been superseded by and the concepts and data have been incorporated within plate tectonics.

By 1915, Alfred Wegener was making serious arguments for the idea in the first edition of The Origin of Continents and Oceans. In that book, he noted how the east coast of South America and the west coast of Africa looked as if they were once attached. Wegener wasn't the first to note this (Abraham Ortelius, Francis Bacon, Benjamin Franklin, Snider-Pellegrini and Frank Bursley Taylor preceded him), but he was the first to marshal significant fossil and paleo-topographical and climatological evidence to support this simple observation (and was supported in this by researchers such as Alex du Toit). However, his ideas were not taken seriously by many geologists, who pointed out that there was no apparent mechanism for continental drift. Specifically, they did not see how continental rock could plow through the much denser rock that makes up oceanic crust. Wegener could not explain the force that propelled continental drift.

Wegener's vindication did not come until after his death in 1930. In 1947, a team of scientists led by Maurice Ewing utilizing the Woods Hole Oceanographic Institution's research vessel Atlantis and an array of instruments, confirmed the existence of a rise in the central Atlantic Ocean, and found that the floor of the seabed beneath the layer of sediments consisted of basalt, not the granite which is the main constituent of continents. They also found that the oceanic crust was much thinner than continental crust. All these new findings raised important and intriguing questions.

Beginning in the 1950s, scientists including Harry Hess, using magnetic instruments (magnetometers) adapted from airborne devices developed during World War II to detect submarines, began recognizing odd magnetic variations across the ocean floor. This finding, though unexpected, was not entirely surprising because it was known that basalt-the iron-rich, volcanic rock making up the ocean floor-contains a strongly magnetic mineral (magnetite) and can locally distort compass readings. This distortion was recognized by Icelandic mariners as early as the late 18th century. More important, because the presence of magnetite gives the basalt measurable magnetic properties, these newly discovered magnetic variations provided another means to study the deep ocean floor. When newly formed rock cools, such magnetic materials recorded the Earth's magnetic field at the time.

As more and more of the seafloor was mapped during the 1950s, the magnetic variations turned out not to be random or isolated occurrences, but instead revealed recognizable patterns. When these magnetic patterns were mapped over a wide region, the ocean floor showed a zebra-like pattern. Alternating stripes of magnetically different rock were laid out in rows on either side of the mid-ocean ridge: one stripe with normal polarity and the adjoining stripe with reversed polarity. The overall pattern, defined by these alternating bands of normally and reversely polarized rock, became known as magnetic striping.

When the rock strata of the tips of separate continents are very similar it suggests that these rocks were formed in the same way implying that they were joined initially. For instance, some parts of Scotland and Ireland contain rocks very similar to those found in Newfoundland and New Brunswick. Furthermore, the Caledonian Mountains of Europe and parts of the Appalachian Mountains of North America are very similar in structure and lithology.

## Floating Continents

The prevailing concept was that there were static shells of strata under the continents. It was observed early that although granite existed on continents, seafloor seemed to be composed of denser basalt. It was apparent that a layer of basalt underlies continental rocks.

However, based upon abnormalities in plumb line deflection by the Andes in Peru, Pierre Bouguer deduced that less-dense mountains must have a downward projection into the denser layer underneath. The concept that mountains had "roots" was confirmed by George B. Airy a hundred years later during study of Himalayan gravitation, and seismic studies detected corresponding density variations.

By the mid-1950s the question remained unresolved of whether mountain roots were clenched in surrounding basalt or were floating like an iceberg. In 1958 the Tasmanian geologist Samuel Warren Carey published an essay The tectonic approach to continental drift in support of the expanding earth model.

## Plate Tectonic Theory

Significant progress was made in the 1960s, and was prompted by a number of discoveries, most notably the Mid-Atlantic ridge. The most notable was the 1962 publication of a paper by American geologist Harry Hess (Robert S. Dietz published the same idea one year earlier in Nature. However, priority belongs to Hess, since he distributed an unpublished manuscript of his 1962 article already in 1960). Hess suggested that instead of continents moving through oceanic crust (as was suggested by continental drift) that an ocean basin and its adjoining continent moved together on the same crustal unit, or plate. In the same year, Robert R. Coats of the U.S. Geological Survey described the main features of island arc subduction in the Aleutian Islands. His paper, though little-noted (and even ridiculed) at the time, has since been called "seminal" and "prescient". In 1967, W. Jason Morgan proposed that the Earth's surface consists of 12 rigid plates that move relative to each other. Two months later, in 1968, Xavier Le Pichon published a complete model based on 6 major plates with their relative motions.

### Explanation of Magnetic Striping

The discovery of magnetic striping and the stripes being symmetrical around the crests of the mid-ocean ridges suggested a relationship. In 1961, scientists began to theorise that mid-ocean ridges mark structurally weak zones where the ocean floor was being ripped in two lengthwise along the ridge crest. New magma from deep within the Earth rises easily through these weak zones and eventually erupts along the crest of the ridges to create new oceanic crust. This process, later called seafloor spreading, operating over many millions of years continues to form new ocean floor all across the 50,000 km-long system of mid-ocean ridges. This hypothesis was supported by several lines of evidence:

1. at or near the crest of the ridge, the rocks are very young, and they become progressively older away from the ridge crest;
2. the youngest rocks at the ridge crest always have present-day (normal) polarity;
3. stripes of rock parallel to the ridge crest alternated in magnetic polarity (normal-reversed-normal, etc.), suggesting that the Earth's magnetic field has reversed many times.

By explaining both the zebralike magnetic striping and the construction of the mid-ocean ridge system, the seafloor spreading hypothesis quickly gained converts and represented another major advance in the development of the plate-tectonics theory. Furthermore, the oceanic crust now came to be appreciated as a natural "tape recording" of the history of the reversals in the Earth's magnetic field.

## Subduction Discovered

A profound consequence of seafloor spreading is that new crust was, and is now, being continually created along the oceanic ridges. This idea found great favor with some scientists, most notably S. Warren Carey, who claimed that the shifting of the continents can be simply explained by a large increase in size of the Earth since its formation. However, this so-called "Expanding earth theory" hypothesis was unsatisfactory because its supporters could offer no convincing mechanism to produce a significant expansion of the Earth. Certainly there is no evidence that the moon has expanded in the past 3 billion years. Still, the question remained: how can new crust be continuously added along the oceanic ridges without increasing the size of the Earth?

This question particularly intrigued Harry Hess, a Princeton University geologist and a Naval Reserve Rear Admiral, and Robert S. Dietz, a scientist with the U.S. Coast and Geodetic Survey who first coined the term seafloor spreading. Dietz and Hess were among the small handful who really understood the broad implications of sea floor spreading. If the Earth's crust was expanding along the oceanic ridges, Hess reasoned, it must be shrinking elsewhere. He suggested that new oceanic crust continuously spreads away from the ridges in a conveyor belt-like motion. Many millions of years later, the oceanic crust eventually descends into the oceanic trenches - very deep, narrow canyons along the rim of the Pacific Ocean basin. According to Hess, the Atlantic Ocean was expanding while the Pacific Ocean was shrinking.

As old oceanic crust is consumed in the trenches, new magma rises and erupts along the spreading ridges to form new crust. In effect, the ocean basins are perpetually being "recycled," with the creation of new crust and the destruction of old oceanic lithosphere occurring simultaneously. Thus, Hess' ideas neatly explained why the Earth does not get bigger with sea floor spreading, why there is so little sediment accumulation on the ocean floor, and why oceanic rocks are much younger than continental rocks.

## Mapping with Earthquakes

During the 20th century, improvements in and greater use of seismic instruments such as seismographs enabled scientists to learn that earthquakes tend to be concentrated in certain areas, most notably along the oceanic trenches and spreading ridges. By the late 1920s, seismologists were beginning to identify several prominent earthquake zones parallel to the trenches that typically were inclined 40-60° from the horizontal and extended several hundred kilometers into the Earth. These zones later became known as Wadati-Benioff zones, or simply Benioff zones, in honor of the seismologists who first recognized them, Kiyoo Wadati of Japan and Hugo Benioff of the United States. The study of global seismicity greatly advanced in the 1960s with the establishment of the Worldwide Standardized Seismograph Network (WWSSN) to monitor the compliance of the 1963 treaty banning above-ground testing of nuclear weapons. The much-improved data from the WWSSN instruments allowed seismologists to map precisely the zones of earthquake concentration world wide.

## Geological Paradigm Shift

The acceptance of the theories of continental drift and sea floor spreading (the two key elements of plate tectonics) may be compared to the Copernican revolution in astronomy (see Nicolaus Copernicus). Within a matter of only several years geophysics and geology in particular were revolutionized. The parallel is striking: just as pre-Copernican astronomy was highly descriptive but still unable to provide explanations for the motions of celestial objects, pre-tectonic plate geological theories described what was observed but struggled to provide any fundamental mechanisms. The problem lay in the question "How?". Before acceptance of plate tectonics, geology in particular was trapped in a "pre-Copernican" box.

However, by comparison to astronomy the geological revolution was much more sudden. What had been rejected for decades by any respectable scientific journal was eagerly accepted within a few short years in the 1960s and 1970s. Any geological description before this had been highly descriptive. All the rocks were described and assorted reasons, sometimes in excruciating detail, were given for why they were where they are. The descriptions are still valid. The reasons, however, today sound much like pre-Copernican astronomy.

One simply has to read the pre-plate descriptions of why the Alps or Himalaya exist to see the difference. In an attempt to answer "how" questions like "How can rocks that are clearly marine in origin exist thousands of meters above sea-level in the Dolomites?", or "How did the convex and concave margins of the Alpine chain form?", any true insight was hidden by complexity that boiled down to technical jargon without much fundamental insight as to the underlying mechanics.

With plate tectonics answers quickly fell into place or a path to the answer became clear. Collisions of converging plates had the force to lift the sea floor to great heights. The cause of marine trenches oddly placed just off island arcs or continents and their associated volcanoes became clear when the processes of subduction at converging plates were understood.

Mysteries were no longer mysteries. Forests of complex and obtuse answers were swept away. Why were there striking parallels in the geology of parts of Africa and South America? Why did Africa and South America look strangely like two pieces that should fit to anyone having done a jigsaw puzzle? Look at some pre-tectonics explanations for complexity. For simplicity and one that explained a great deal more look at plate tectonics.

A great rift, similar to the Great Rift Valley in northeastern Africa, had split apart a single continent, eventually forming the Atlantic Ocean, and the forces were still at work in the Mid-Atlantic Ridge.

We have inherited some of the old terminology, but the underlying concept is as radical and simple as was "The Earth moves" in astronomy.

## Biogeographic Implications on Fauna and Flora

Continental drift theory helps biogeographers to explain the disjunct biogeographic distribution of present day plants and animals found on different continents but having similar ancestors. In particular, it explains the Gondwanan distribution of ratites and the Antarctic flora.

## Plate Tectonics on other Planets

- *Mars:* As a result of 1999 observations of the magnetic fields on Mars by the Mars Global Surveyor spacecraft, it has been proposed that the mechanisms of plate tectonics may once have been active on the planet - see Geology of Mars. Further data from the Mars Express orbiter's High Resolution Stereo Camera in 2007 clearly showed an example in the Aeolis Mensae region.
- *Venus:* Venus shows no evidence of active plate tectonics. There is debatable evidence of active tectonics in the planet's distant past; however, events taking place since then (such as the plausible and generally accepted hypothesis that the Venusian lithosphere has thickened greatly over the course of several hundred million years) has made constraining the course of its geologic record difficult. However, the numerous well-preserved impact craters has been utilized as a dating method to approximately date the Venusian surface (since there are thus far no known samples of Venusian rock to be dated by more reliable methods). Dates derived are the dominantly in the range ~500 Mya - 750 Mya, although ages of up to ~1.2 Gya have been calculated. This research has led to the fairly well accepted hypothesis that Venus has undergone an essentially complete volcanic resurfacing at least once in its distant past, with the last event taking place approximately within the range of estimated surface ages. While the mechanism of such an impressionable thermal event remains a debated issue in Venusian geosciences, some scientists are advocates of processes involving plate motion to some extent.

- *Galilean Satellites:* Some of the satellites of Jupiter have features that may be related to plate-tectonic style deformation, although the materials and specific mechanisms may be different from plate-tectonic activity on Earth.

## Metaphoric Uses

Sometimes the idea of moving tectonic plates is used metaphorically, e.g. "a tectonic shift" in a BBC TV news program describing the political effects of Ariel Sharon's illness on 4 January 2005.

In the late 1980s, Québec theatre director Robert Lepage created a large international production called Tectonic Plates, which used this image to illustrate the rifts between Europe and America and the drifting of various destinies, relative to one another.

# PALEONTOLOGY

Paleontology, palaeontology or palæontology (from Greek: paleo, "ancient"; ontos, "being"; and, logos, "knowledge") is the study of prehistoric life forms on Earth through the examination of plant and animal fossils. This includes the study of body fossils, tracks (ichnites), burrows, cast-off parts, fossilised faeces (coprolites), palynomorphs and chemical residues. See also paleoanthropology.

## Background

Modern paleontology sets ancient life in its contexts by studying how long-term physical changes of global geography paleogeography and climate paleoclimate have affected the evolution of life, how ecosystems have responded to these changes and have adapted the planetary environment in turn and how these mutual responses have affected today's patterns of biodiversity.

Hence, paleontology overlaps with geology (the study of rocks and rock formations) as well as with botany, biology, zoology and ecology - fields concerned with life forms and how they interact.

The major subdivisions of paleontology include paleozoology (animals), paleobotany (plants) and micropaleontology (microfossils). Paleozoologists may specialise in invertebrate paleontology, which deals with animals without backbones or in vertebrate paleontology, dealing with fossils of animals with backbones, including fossil hominids (paleoanthropology). Micropaleontologists study microscopic fossils, including organic-walled microfossils whose study is called palynology.

There are many developing specialties such as paleobiology, paleoecology, ichnology (the study of tracks and burrows) and taphonomy (the study of what happens to organisms after they expire). Major areas of study include the correlation of rock strata with their geologic ages and the study of evolution of lifeforms.

Paleontology utilises the same classic binomial nomenclature scheme, devised for the biology of living things by the mid-18th century Swedish biologist Carolus Linnaeus and increasingly sets these species in a genealogical framework, showing their degrees of interrelatedness using the still somewhat controversial technique of 'cladistics'.

The primary economic importance of paleontology lies in the use of fossils to determine the age and nature of the rocks that contain them or the layers above or below. This information is vital to the mining industry and especially the petroleum industry. Simply looking at the fossils contained in a rock remains one of the fastest and most accurate means of telling how old that rock is.

Fossils were known by primitive humans and were sometimes identified correctly as the remains of ancient lifeforms. The organised study of paleontology dates from the late 18th century. For a more complete historical overview see the article History of paleontology.

**Notable Paleontologists**

History includes a number of prominent paleontologists. Charles Darwin collected fossils of South American mammals during his trip on the Beagle and examined petrified forests in Patagonia. Mary Anning was a notable early paleontologist.

She found several landmark fossils, in her home town of Lyme Regis. Although self-taught, she collected and described them in a very systematic way. William Buckland, Richard Owen, Gideon Mantell, Georges Cuvier and Thomas Huxley were important early pioneers, in the field of paleontology. Thomas Jefferson took a keen interest in mammoth bones. Edward Drinker Cope and Othniel Charles Marsh waged a famously fierce competition known as the Bone Wars in the late 19th century that involved some questionable practices, but which significantly advanced the understanding of the natural history of North America and vertebrate paleontology.

Professor Earl Douglass of the Carnegie University in Pittsburgh, Pensylvannia, opened the fossil quarry protected today by Dinosaur National Monument in Utah. Douglass' fossils are in several Natural History Museums. Meanwhile, Baron Franz Nopcsa, a pioneer paleobiologist, argued that dinosaurs might have been both warm-blooded and ancestral to birds.

Besides looking at mammal teeth and unearthing penguin skeletons, George Gaylord Simpson played a crucial role in bringing together ideas from biology, paleontology and genetics, to help create the 'Modern Synthesis' of evolutionary biology. His book "Tempo and Mode" is a classic in the field, Prominent names in invertebrate paleontology include Steven Stanley, Stephen Jay Gould, David Raup, Rousseau H. Flower and Jack Sepkoski, who have done much to expand our understanding of long-term patterns in the evolution of life on earth. Large names in the field of paleoanthropology include Louis, Mary and Richard Leakey, Raymond Dart, Robert Broom, C.K. 'Bob' Brain, Kenneth Oakley, Robert Ardrey and Donald Johanson. In recent times, Mongolian paleontologist Rinchen Barsbold has done much to expand our understanding of dinosaur and bird evolution. Paul Sereno of the Field Museum in Chicago has made several important dinosaur finds in areas such as the Sahara, where fossil hunting has been uncommon.

**Human Geography**

Human geography is a branch of geography that focuses on the study of patterns and processes that shape human interaction with various environments. It encompasses human, political, cultural, social, and economic aspects. While the major focus of human geography is not the physical landscape of the Earth (see physical geography), it is hardly possible to discuss human geography without referring to the physical landscape on which human activities are being played out, and environmental geography is emerging as a link between the two. Human geography can be divided into many broad categories, such as:

## ECONOMIC GEOGRAPHY

Economic geography is the study of the location, distribution and spatial organisation of economic activities across the Earth. It focuses on the location of industries and retail and wholesale businesses, on transportation and trade, and on the changing value of real estate. Courses in economic geography may cover such topics as transportation, agriculture, industrial location, world trade, and the spatial organisation and function of business activity.

### Areas of Study

The distribution of economic activities on Earth is influenced by many environmental, social, political, historical and other factors. Geology can affect resource availability, geomorphology, the cost of transportation, and the quality of soiled land alter economic activities. Climate can influence natural resource availability (forestry products) and location or type of agriculture (see growing region). The social and political factors that are unique to a particular region also have an impact on economic decisions and further distributions of these activities.

Economic geography research focuses on the study of spatial aspects of economic activities on various scales. The distance to the city (or Central business district) as a marketplace with demand for products plays a significant role in economic decisions of firms while other factors such as access to the sea and the presence of raw materials like oil affects the economic conditions of countries. Singapore, for example, occupies a key position as a seaport, while the wealth of Saudi Arabia depends almost entirely on oil.

In today's world location, distribution and character of economic activities is much influenced by globalisation. States and their borders play less significant role as many countries tend to eliminate the effects of borders and deepen the mutual cooperation on the global scale. Border regions that are often economically marginal and underdeveloped are also better cooperating with each other. The best example is the creation of European Union. Significant characteristics is also the occurrence of large business clusters that are forming around the world.

## Approaches to Study

- Theoretical economic geography focuses on building theories about spatial arrangement and distribution of economic activities.
- Historical economic geography examines history and the development of spatial economic structure.
- Regional economic geography examines the economic conditions of particular regions or countries of the world. It deals with economic regionalisation as well.
- Critical economic geography is approach from the point of view of contemporary critical geography and its philosophy.
  * Behavioral economic geography examines the cognitive processes underlying spatial reasoning, locational decision making, and behavior of firms and individuals.

## Subdivision

Thematically economic geography can be divided into these subdisciplines:

- Geography of agriculture
- Geography of industry
- Internet Geography
- Geography of services
- Geography of transportation
- and others

However, their areas of study may overlap with another geographical sciences or may be considered on their own.

### History of Economic Geography

In the history of economic geography there were many influences coming mainly from economics and geographical sciences.

First traces of the study of spatial aspects of economic activities on Earth can be found in Strabo's Geographika written almost 2000 years ago.

During the period known in geography as environmental determinism notable (though later much criticized) influence came from Ellsworth Huntington and his theory of climatic determinism.

Valuable contributions came from location theorists such as Johann Heinrich von Thünen or Alfred Weber. Other influential theories were Walter Christaller's Central place theory, the theory of core and periphery.

Big impact on economic geography had the Fred K. Schaefer's article Exceptionalism in geography: A Methodological Examination published in American journal Annals (Association of American Geographers) and his critique of regionalism. The article became a rallying point for the younger generation of economic geographers who were intent on reinventing the discipline as a science. Quantitative methods became prevailing in research. Well-known economic geographers of this period are William Garrison, Brian Berry, Waldo Tobler, Peter Haggett, William Bunge and others.

Contemporary economic geographers tend to specialize in areas such as location theory and spatial analysis (with the help of geographic information systems), market research, geography of transportation, land or real estate price evaluation, regional and global development, planning, Internet geography, and others.

## TRANSPORTATION GEOGRAPHY

Transportation geography is the branch of geography that describes the spatial aspects of interactions between humans and their use of vehicles or other modes of travelling. It is a branch of Urban geography.

The links between transport and urban geography are closely intertwined, with the concept of ribbon development being closely aligned to urban and transport studies. As humans increasingly seek to travel the world, the relationship between transport and urban areas have often become obscured.

Transportation geography measures the result of human activity between and within locations. It focuses on items such as travel time, routes undertaken, modes of transport, resource use and sustainability of transport types on the natural environment. Other sections consider topography, safety aspects of vehicle use and energy use within an individual's or group's journey.

## DEVELOPMENT GEOGRAPHY

Development geography is the study of the Earth's geography and its relationship with economic development. It is very closely related to economic geography and development economics. This branch of human geography is intertwined with resource distribution and consumption, demographics, soil quality, topography, climate and natural disasters.

As the world is divided into the More Economically Developed Countries (MEDCs, also called Developed countries and First World Countries) and Less Economically Developed Countries (LEDCs, also Developing countries and Third World Countries) wealth (both money and materials resources) is unequally distributed among the world's population with large consequences for people's lifestyles and the environment. In development geography geographers study spatial patterns in development and try to find by what characteristics they can measure economic development. They seek to understand both the geographical causes and consequences of varying development.

### Measuring Development

There are many methods that geographers use to quantify and compare the development of different countries and they have different strengths and weaknesses.

Quantitative Indicators are numerical indications of development.

- GNP per capita measures the value of all the goods and services produced in a country, excluding those produced by foreign companies. This figure is widely used but has many problems. It does not take into account the distribution of the money which can often be extreme unequal as in the UAE where oil money has been collected by a rich elite and has not flowed to the bulk of the country. GNP does not measure whether the money produced is actually improving people's lives and this is important because in many MEDCs where there are large increases in wealth over time but only small increases in happiness. The figure rarely takes into account the unofficial economy, which includes subsistence agriculture and cash-in-hand or unpaid work, which is often substantial in LEDCs. In LEDCs it is often too expensive to accurately collect this data and some governments intentionally or unintentionally release inaccurate figures. The figure is usually given in US dollars which due to changing currency exchange rates can distort the money's true street value so it is often converted using purchasing power parity (PPP) in which the actual comparative purchasing power of the money in the country is calculated.
- Social (demographic) indicators include the birth rate, death rate and fertility rate.
- Health indicators include nutrition (calories per day, calories from protein, percentage with malnutrition), infant mortality and population per doctor.
- Economic indicators include unemployment rates, energy consumption and percentage of GNP in primary industries.

Composite Indicators combine several quantitative indicators into one figure and generally provide a more balanced view of a country. Usually they include one economic, one health and one educational indicator.

- The HDI (Human Development Index) is now the most widely used composite indicator. A number is calculated between 0 and 1 taking into account the most important measures: GNP per capita, the adult literacy rate, the school enrollment rate and life expectancy. It was started by the United Nations in 1990 to replace GNP as a more accurate way of measuring development. A HDI between 1 and 0.8 is considered high, 0.8 and 0.6 is considered medium and 0.6 to 0.4 is considered low.
- Other composite measures include the PQLI (Physical Quality of Life Index) which was a precursor to the HDI which used infant mortality rate instead of GNP per capita and rated countries from 0 to 100. It was calculated by assigning each country a score of 0 to 100 for each indicator compared with other countries in the world. The average of these three numbers makes the PQLI of a country.
- The HPI (Human Poverty Index) is used to calculate the percentage of people in a country who live in relative poverty. In order to better differentiate the number of people in abnormally poor living conditions the HPI-1 is used in developing countries, and the HPI-2 is used in developed countries. The HPI-1 is calculated based on the percentage of people not expected to survive to 40, the adult illiteracy rate, the percentage of people without access to safe water, health services and the percentage of children under 5 who are underweight. The HPI-2 is calculated based on the percentage of people who do not survive to 60, the adult functional illiteracy rate and the percentage of people living below 50% of median personal disposable income.
- The GDI (Gender-related Development Index) measures gender equality in a country in terms of life expectancy, literacy rates, school attendance and income.

Qualitative Indicators include descriptions of living conditions and people's quality of life. They are useful in analysing features that are not easily converted to numbers such as freedom and security.

***Data Example***

| HDI rank | Country | GDP per capita (PPP US$) (HDI) 2002 | Human development index (HDI) value 2002 |
|---|---|---|---|
| 3 | Australia | 28,260 | 0.946 |
| 72 | Brazil | 7,770 | 0.775 |
| 147 | Zimbabwe | 2,400 | 0.491 |

**Source:** UN Human Development Report (HDR). The UN Human Development Report is the standard for most development statistics worldwide and provides them freely on their website.

## Geographic Variations in Development

It is important to grasp just how unequally distributed the world's wealth is. Economic growth over the past 50 years has been impressive. In general richer countries are improving at a faster rate than the poorer countries because they have more money to invest in improvements.

> *"Global wealth also increased in material terms, and during the period 1947 to 2000, average per capita incomes tripled as global GDP increased almost tenfold (from $US3 trillion to $US30 trillion)... Over 25% of the 4.5 billion people in LEDCs still have life expectancies below 40 years. More than 80 countries have a lower annual per capita income in 2000 than they did in 1990. The average income in the world's five richest countries is 74 times the level in the world's poorest five, the widest it has ever been. Nearly 1.3 billion people have no access to clean water. About 840 million people are malnourished." - Codrington, Stephen. Planet Geography 3rd Edition (2005) Page 97*

The most famous pattern in development is the North-South divide. The North-South divide is the divide which separates the rich North or the developed world, from the poor South. This line of division is not as straightforward as it sounds and splits the globe into two main parts.

The "North" in this divide is regarded as being North America, Europe, Russia, Japan, Australia and New Zealand.

The countries within this area are generally the more economically developed.

The "South" therefore encompasses the remainder of the Southern Hemisphere, mostly consisting of LEDCs. Another possible dividing line is the Tropic of Cancer with the exceptions of Australia and New Zealand.

It is critical to understand that the status of countries is far from static and the pattern is likely to become distorted with the fast development of certain southern countries, many of them NICs (Newly Industrialised Countries) including Thailand, Brazil, Malaysia, Mexico and others. These countries are experiencing sustained fast development on the back of growing manufacturing industries and exports.

Most countries are experiencing significant increases in wealth and standard of living. However there are unfortunate exceptions to this rule. Noticeably the former Soviet Union has experienced major disruption of industry in the transition to a market economy. Many African nations have recently experienced reduced GNPs due to wars and the AIDS epidemic, including Angola, Congo, Sierra Leone and others.

Arab oil producers rely very heavily on oil exports to support their GDPs so any reduction in oil's market price (currently unlikely) can lead to rapid decreases in GNP. Countries which rely on only a few exports for much of their income are very vulnerable to changes in the market value of those commodities and are often derogatively called banana republics.

Many developing countries do rely on exports of a few primary goods for a large amount of their income (coffee and timber for example), and this can create havoc when the value of these commodities drops, leaving these countries with no way to pay off their debts.

Within countries the pattern is that wealth is more concentrated around urban areas than rural areas. Wealth also tends towards areas with natural resources or in areas that are involved in tertiary (service) industries and trade. This leads to a gathering of wealth around mines and monetary centres such as New York, London and Tokyo.

## Causes of Inequality

There are many reasons why some countries develop faster than others. They can be placed under 5 headings using the mnemonic SHEEP, but there is much overlap:

### Social

The more money a country has, the more it can spend on health care, education and birth control. Social traditions that discourage birth control increase birth rates and impede the economic development of poor countries.

Different societies value hard work, material gain and social cohesion differently and this will clearly have an effect on growth and efficiency. Also, there is the vicious "poverty cycle." This is when the general attitude in developing countries is to have large amounts of children, as they are a guarantee, almost insurance for later life.

However, it is hard to cope with large amounts of children, this leads to over population in a country and therefore more poverty. The cycle then starts again.

### Historical

Historically colonialism has probably had the largest influence on development. It channeled resource wealth towards Europe and North America at the expense of many African, South American and Asian countries which did not receive reasonable prices for their goods.

European colonizers build strong industries from this wealth while not investing in such development in their colonies. At the end of colonialism many countries were left without the social, economic or political structures that encouraged development so poverty became entrenched.

In many cases artificial borders were drawn which did not reflect the desires of the local inhabitants, leading to civil wars or social instability. Other historical influences can include incompetent governments or a retention of tribal lifestyles that prevented countries from developing economically.

## Economic

Countries with resources such as iron ore, oil and coal are likely to develop industrially more easily because they do not have to import these resources, leading to debt. Their extraction and sale create jobs and transport systems while giving certain countries trade and political leverage over others.

The resources can also earn large sums of money in trade, allowing a country to invest in other industries. Many European nations developed during the industrial revolution on the back of coal and iron industries.

However, the fact that many resource-rich (oil in particular) African and Middle-Eastern nations have not developed economically while their resources are mined demonstrates that these factors are not sufficient in themselves. Often kleptocracies develop around these industries and grow very rich while investing little in the country's population itself. Nigeria and Saudi Arabia are two examples of this. In fact the wealth generated can often help to entrench an incompetent dictatorship in power or even lead to destructive resource wars as has occurred in Africa.

## Environmental

Natural hazards including flooding, droughts, earthquakes, volcanoes, storms, hurricanes, diseases, illnesses and pests all prevent economic development. Large natural disasters can set countries back greatly in their economic development, as in periodic flooding of Bangladesh. Volcanos and floods can often have both positive and negative effects as they bring in nutrient rich sediment.

Areas around volcanoes and flooding deltas are often heavily populated, as in Egypt, Bangladesh and Indonesia. Diseases such as malaria which thrive in tropical climates and AIDS which is endemic in Africa prevent people from working and create an economic burden on society. Pests such as locusts reduce agricultural output and make it more difficult for a country to earn sufficient money to escape from subsistence agriculture. Reliable sources of water are necessary for productive agriculture and to a lesser extent industry. Human induced environmental problems include desertification, salinity, water pollution, land clearing and many more. Desertification is caused by poor land management removing the nutrients necessary for plant growth.

It is a worldwide problem with massive consequences for the countries it affects. Salinity is caused by poor irrigation techniques. Water pollution from industry can include acids and bases, poisonous minerals and material with a high BOD which cause algal blooms. This pollution makes it more difficult for a populations to access fresh water. Logging initially brings in investment but often land with trees removed is of far reduced agricultural value and is vulnerable to desertification. Logged rain forests are especially vulnerable to mineral leeching due to high rainfall and often become worthless. As tourism is now a major source of income for most LEDCs it is necessary to care for natural resources which can bring in this long-term source of wealth.

# 5

# POPULATION GEOGRAPHY

## INTRODUCTION

Population Geography is a division of Human Geography. It is the study of the ways in which spatial variations in the distribution, composition, migration, and growth of populations are related to the nature of places.

Population geography involves demography in a geographical perspective. It focuses on the characteristics of population distributions that change in a spatial context. Examples can be shown through population density maps. A few types of maps that show the spatial layout of population are chloropleth, isoline, and dot maps. Demography studies:

- Study of people in their spatial distribution and density
- Increase or decrease in population numbers
- The movements and mobility of populations
- Occupational Structure
- Grouping of people in settlements
- The way from the geographical character of places e.g. settlement patterns
- The way in which places in turn react to population phenomena e.g. immigration

All of the above are looked at over space and time.

**N.B.:** The boundary between population geography and demography is becoming more and more blurred.

## URBAN GEOGRAPHY

Urban geography is the study of urban areas. That is the study of areas which have a high concentration of buildings and infrastructure. These are areas where the majority of economic activities are in the secondary sector and tertiary sectors. They probably have a high population density.

It can be considered a part of the larger field of human geography. However, it can often overlap with other fields such as anthropology and sociology. Urban geographers seek to understand how factors interact over space, what function they serve and their interrelationships. Urban geographers also look at the development of settlements. Therefore, it involves planning city expansion and improvements. Urban geography, then, attempts to account for the human and environmental impacts of the change.

Urban geography forms the theoretical basis for a number of professions including urban planning, retail store site selection, real estate development, crime pattern analysis and logistical analysis.

### Areas of Study

There are essentially two approaches to urban geography. The study of problems relating to the spatial distribution of cities themselves and the complex patterns of movement, flows and linkages that bind them in space. Studies in this category are concerned with the city system. Secondly there is the study of patterns of distribution and interaction within cities, essentially the study of their inner structure. Studies in this category are concerned with the city as a system. A succinct way to define urban geography that recognizes the link between these two approaches within the subject is then, that "urban geography is the study of cities as systems within a system of cities."

### Site and Situation

Site describes the location of a city with respect to its soil, water supply and relief, or more still the actual point on which a settlement is built while situation describes the surrounds of the city in terms of other settlements, rivers, mountains and communication. Locations for cities are usually chosen for good reasons. Benefits of certain locations can include:

- A wet area: water is a constant necessity for urban areas and is difficult to transport. For this reason many cities are located near or adjacent to rivers.
- A dry area: in wet areas a dry area offers protection from flooding and marshland.
- Easy access to building materials: stone, wood or clay are necessary for the construction of cities and are difficult to transport long distances.
- A strategic defensive position: historically many cities have been constructed on high ground in order to make attack more difficult and to give a good view of surrounding land. River meanders are also used as partial moats.
- fuel supply: most cities were initially constructed near wood for burning and cooking. Today many cities are constructed near coal, oil and gas mines to make use of those resources.
- A food supply: cities need some nearby land to be suitable for animal grazing or crop growing
- A travel intersection point and bridging points: it is often useful for a city to be located at the intersection of rivers, roads or train lines in order to facilitate travel and trade. Bridging points are shallow areas that allow easy construction of bridges.
- Shelter and aspect: it is desirable to construct cities located on the side of a slope that is protected from incoming winds, and in a direction that receives maximum sun exposure.

## The City System

Much of the world is increasingly urbanized. Before 1850 no country could claim to be predominantly urbanized and at the turn of the century, only Great Britain could be regarded as such.

Urbanization is the process whereby society is transformed from an essentially rural one to a predominantly urban one. Its most visible expression in the landscape is the growth of cities and an increase in their number, size and importance. Also, urbanization process can be closely allied with economic colonization.

Urbanization, however is not just the growth of cities but rather it is a complex change in economic, social and political thinking. As cities grow, new definitions have been introduced for census purposes in which the concept of the "extended city" is commonly used. Definitions such as the urban area based on the physical extent of the built-up area and minimum requirements of population size and density as well as the metropolitan area, based on urban population size and commute patterns as a measure of spatial integration give us a much better indication of the population size of cities than the use of political entities like the municipality.

One of the definitions that have raised much interest is that based on the concept of urban field which is a new form of urban habitat of relatively low density involving a good transportation system and a broad array of economic, social and recreational opportunities. Each urban field is centered on and dominated to a certain extent by a metropolitan area of at least 200,000-300,000 people. It's outer limits can be defined by two criteria:

(1) The maximum time or distance that most people are prepared to commute.

(2) The time or distance that most people are prepared to spend traveling to or from weekly or weekend recreational activities. Many of these urban fields come into contact with each other and they interact.

Cities do not and cannot exist in isolation. Whether you view the developments in transportation and communication as prerequisites or agents for the growth of cities, the concentration of surplus products in cities necessitated linkages initially with their surrounding regions and later as settlement proceeded westward and a spatial division of labour accompanied the geographic specialization of production, with other regions and cities. In order for the surplus to be concentrated at particular locations linkages were necessary and cities became nodes in networks comprising of the movement of goods, services, materials, people, money, credit, investment and information. These various forms of flows, movements, transactions and linkages are collectively referred to as spatial interaction and acts as the key to the city system.

Interaction then plays a number of crucial roles in shaping the form and structure of city systems, as it does in the internal structure of cities. Four roles the city system plays are particularly important. Firstly, in the same way that the market economy is integrated through price-fixing mechanisms, so interaction performs a spatial integrating role. Second, interaction permits differentiation of functional specialization of cities within the city system. Third and most important, interaction is the medium of spatial organization. Lastly, interaction is extremely important in bringing about change and the reorganization of spatial relationships within the city system.

The diffusion of innovations is the ideal example of the function of the urban hierarchy. Large cities are more the most likely place for invention and innovation to occur and spread or diffuse to other cities below it in the hierarchy. As some cities grew and become more important they climb the hierarchy, with major metropolitan centers at the top and towns, villages and hamlets at the bottom.

## Cities as Centers of Manufacturing and Services

Cities differ in their economic makeup, their social and demographic characteristics and the roles they play within the city system. These differences can be traced back to regional variations in the local resources on which, growth was based during the early development of the urban pattern and in part the subsequent shifts in the competitive advantage of regions brought about by changing locational forces affecting regional specialization within the framework of the market economy. Recognition of different city types necessitates their classification, and it is to this important aspect of urban geography that we now turn. Emphasis is on functional town classification and the basic underlying dimensions of the city system.

The purpose of classifying cities is twofold. On the one hand, it is undertaken in order to search reality for hypotheses. In this context, the recognition of different types of cities on the basis of, for example, their functional specialization may enable the identification of spatial regularities in the distribution and structure of urban functions and the formulation of hypotheses about the resulting patterns. On the other hand, classification is undertaken to structure reality in order to test specific hypotheses that have already been formulated. For example, to test the hypotheses that cities with a diversified economy grow at a faster rate then those with a more specialized economic base, cities must first be classified so that diversified and specialized cities can be differentiated.

The simplest way to classify cities is to identify the distinctive role they play in the city system. There are three distinct roles. 1. Central places functioning primarily as service centers for local hinterlands. 2. Transportation cities performing break-of-bulk and allied functions for larger regions. 3. Specialized-function cities are dominated by one activity such as mining, manufacturing or recreation and serving national and international markets. The composition of a cities labor force has traditionally been regarded as the best indicator of functional specialization, and different city types have been most frequently identified from the analysis of employment profiles. Specialization in a given activity is said to exist when employment in it exceeds some critical level.

The relationship between the city system and the development of manufacturing has become very apparent. The rapid growth and spread of cities within the heartland-hinterland framework after 1870 was conditioned to a large extent by industrial developments and that the decentralization of population within the urban system in recent years is related in large part to the movement of employment in manufacturing away from the traditional industrial centers. Manufacturing is found in nearly all cities, but its importance is measured by the proportion of total earnings received by the inhabitants of an urban area. When 25 percent or more of the total earnings in an urban region are derived from manufacturing, that urban areas is arbitrarily designated as a manufacturing center.

The location of manufacturing is affected by myriad economic and non-economic factors, such as the nature of the material inputs, the factors of production, the market and transportation costs. Other important influences include agglomeration and external economies, public policy and personal preferences. Although it is difficult to evaluate precisely the effect of the market on the location of manufacturing activities, two considerations are involved: the nature of and demand for the product and transportation costs.

## SOCIAL GEOGRAPHY

Social geography is the study of how society affects geographical features and how environmental factors affect society.

### Case Study: India

Victims of their own historical success, Indians suffer from a rural economy. The reason? A high population density, poverty and strong echoes of the traditional caste system holds back any progress or urbanization. The fertile Ganges Valley with monsoon rain and river always supported a dense rural population. Rice is the stable crop. A settled traditional agriculture is practised on small plots, but tenants are exploited by landlords. There is a large mass of landless labourers. Poverty still acute, however, the emerging middle class peasantry benefited from the Green Revolution.

### Areas of Study

Questions in the field of social geography might include the examination of rural exodus or urban exodus or whether low-rise developments generate a different type of daily life than tower blocks.' It deals also with problems of segregation and discrimination, socio-spatial variations in health, analysis of spatial crime patterns and others.

In the field of community development (or community economic development), the importance of place has been a focal point for sociologists to determine what effects geography may have on a local community's cohesiveness and the sense of community. Studies in community psychology suggest that where we are many times has an effect on who we are.

## BEHAVIORAL GEOGRAPHY

Behavioral geography is an approach to Human Geography that examines human behavior using a disaggregate approach. Behavioral Geographers focus on the cognitive processes underlying spatial reasoning, decision making, and behavior.

Because of the name it is often incorrectly assumed to have its roots in behaviourism. Due to the emphasis on cognition, this is clearly not the case.

The cognitive processes include environmental perception and cognition, wayfinding, the construction of cognitive maps, place attachment, the development of attitudes about space and place, decisions and behavior based on imperfect knowledge of one's environs, and numerous other topics.

The approach adopted in behavioral geography is closely related to that of psychology, but draws on research findings from a multitude of other disciplines including economics, sociology, anthropology, transportation planning, and many others.

### Geosophy

Geosophy is a concept introduced to geography by J.K. Wright in 1947. The word is a compound of 'ge' (Greek for earth) and 'sophia' (Greek for knowledge). Wright defined it thus:

> *Geosophy... is the study of geographical knowledge from any or all points of view. It is to geography what historiography is to history; it deals with the nature and expression of geographical knowledge both past and present-with what Whittlesey has called 'man's sense of [terrestrial] space'. Thus it extends far beyond the core area of scientific geographical knowledge or of geographical knowledge as otherwise systematized by geographers. Taking into account the whole peripheral*

*realm, it covers the geographical ideas, both true and false, of all manner of people-not only geographers, but farmers and fishermen, business executives and poets, novelists and painters, Bedouins and Hottentots-and for this reason it necessarily has to do in large degree with subjective conceptions. (Wright 1947)*

This has been summarised as:

*the study of the world as people conceive of and imagine it (McGreevy 1987)*

*Belief systems as they relate to human interaction with the Earth's environments. (attributed to Professor Innes Park 1995)*

## Cultural Region

Cultural region is a term used mainly in the study of geography. Distinct cultures often do not limit their geographic coverage inside the borders of a nation state, or to smaller subdivisions of a state. To 'map' a culture, we often have to identify an actual 'cultural region', and when we do this we find that it bears little relationship to the legal borders drawn up by custom, treaties, charters or wars.

There are different kinds of cultural regions that can be delineated. A map of culture that maps 'religion & folklore' may have slightly different shape to one which, in the same region, maps 'dress and architecture'.

## TYPES OF CULTURAL REGION

### Cultural Regions of the World

#### *Anglo-America*

Anglo-America is a term used to describe those parts of the Americas in which English is the main language, or having significant historical, ethnic, linguistic, and cultural links to England/United Kingdom or the British Isles in general. Alternatively, Anglo-America is the American portion of the Anglosphere. Anglo-America is distinct from Latin America, a region of the Americas where Romance languages derived from Latin (namely, French, Spanish and Portuguese) are prevalent.

Anglo-America includes the United States and Canada in North America, and the term is frequently used in reference to the two countries together. In Middle and South America, Belize, Guyana, Jamaica, and several other Caribbean territories may also be included, as is Bermuda (a British possession 1 000 kilometres east of the American mainland); when referring to this broader group, the term Anglophone America is sometimes used. Suriname is not a part of Anglo-America because Dutch is the official language there, like in the Netherlands Antilles and Aruba. English is also the official language of the Falkland Islands.

The adjective Anglo-American is used in the following ways:

- to denote the cultural sphere shared by the United Kingdom, the United States, and sometimes English Canada. For example, "Anglo-American culture is different from French culture." Political leaders including Sir Winston Churchill, Franklin Roosevelt, and Ronald Reagan have utilized the term to discuss the "special relationship" between the United States and the United Kingdom.
- to describe relations between the United Kingdom on one hand and the Americas, in particular the United States, on the other. For example, "Anglo-American relations were tense before the War of 1812."

As a noun, Anglo-American can refer to an English speaking European American, sometimes shortened to Anglo. This usage occurs most frequently in the discussion of the history of English-speaking people of the United States and the Spanish-speaking people residing in the western U.S. during the Mexican-American War. This usage generally ignores the distinctions between English Americans, German Americans, Irish Americans, and other northern European descent peoples, comprising the majority of English-speaking Europeans in the United States.

**Austronesia**

Austronesia, in historical terms, refers to the homeland of the people who speak Austronesian languages, to which Malay,

Filipino, Indonesian, Malagasy, native Hawaiian, the Fijian language and around a thousand other languages belong. The Austronesian homeland is thought by linguists to have been prehistoric Taiwan.

The name Austronesia comes from the Latin australis "southern" plus the Greek "island".

However, in present terminologies, the word Austronesia pertains to the regions where Austronesian languages are spoken. Austronesia then covers about half of the globe, although mostly ocean, and oceanic islands, starting from Madagascar to the west until Easter Island, to the east. Austronesia as a region has three traditional divisions: Formosa (Taiwan), Nusantara (the Malay Archipelago), and Austronesian Oceania (Micronesia, Melanesia, and Polynesia).

## Germanic Europe

Germanic Europe is the part of Europe in which Germanic languages are predominant. Countries or areas in which such language is officially recognized and/or de facto spoken as a minority language are sometimes included; this entire area corresponds more or less to North-Western Europe and western parts of Central Europe.

In its widest sense, this region consists of Iceland, Ireland, the United Kingdom, the Faroe Islands, the Netherlands, Denmark, Sweden, Norway, the Swedish-speaking municipalities of Finland, French Flanders and Alsace-Moselle in France, Flanders and the smaller German-speaking Community in Belgium, the German-speaking part of Luxembourg, Germany, the formerly German parts of Poland as well as in East Prussia and the Baltic States Estonia and Latvia, Liechtenstein, the German-speaking part of Switzerland, Austria, and the South Tyrol autonomous province in Italy.

At least parts of Scotland, the Isle of Man, Wales, Cornwall and Northern Ireland in the UK as well as the Republic of Ireland belong historically and culturally mainly to Celtic Europe.

The predominant religion in the majority of the region is Protestantism; the national churches of the United Kingdom, the Nordic countries, the Netherlands, northern Germany, and parts of Switzerland are Protestant. At the same time, some parts of the region are Catholic: the Republic of Ireland, southern Germany (particularly Bavaria), Austria, Belgium, the southern regions of the Netherlands, and central and southern Switzerland.

### Slavic Europe

Slavic Europe is a region of Europe where Slavic languages are spoken. This area corresponds, more or less, to Central and Eastern Europe and the Balkans, and consists of: Belarus, Bosnia and Herzegovina, Bulgaria, Croatia, the Czech Republic, the Republic of Macedonia, Montenegro, Poland, Russia, Serbia, Slovakia, Slovenia, the disputed territory of Transnistria, and Ukraine.

The main religions are Orthodox Christianity and Catholicism, with large Muslim populations in some parts formerly ruled by the Ottoman Empire, as well as various constituent parts of Russia. There were also over 4 million Jews living in Slavic Europe until the Holocaust greatly reduced that number and many surviving Jews emigrated to nearby Israel and the United States.

There were a large amount of Jews in Slavic Eastern Europe because many Jews had been expelled during the Middle Ages from anti-Semitic, Christian, Western Europe. (See Christianity and anti-Semitism).

Amongst the total Slavic population of Europe, there are local Slavic minorities of Sorbs in Germany and Lipovans in Romania, as well as a Slovenian and Croatian minorities in Northeastern Italy and Southern Austria, and the significant Slavic diaspora in Western Europe.

### Latin America

Latin America (Portuguese: América Latina, Spanish: América Latina, French: Amérique Latine) is the region of the Americas where Romance languages, those derived from Latin (particularly Spanish and Portuguese), are primarily spoken. Latin America is contrasted with Anglo-America, that region of the Americas where English predominates.

## Definition

There are several definitions of Latin America, none of them perfect or necessarily logically consistent:

- In most common contemporary usage, Latin America refers only to those territories in the Americas where Spanish or Portuguese prevail: Mexico, most of Central and South America, plus Cuba, Puerto Rico, and the Dominican Republic in the Caribbean.
- However, strictly speaking, Latin America could designate all those countries and territories in the Americas where Romance languages - Spanish, Portuguese, but also French, and their creoles - are spoken. Indeed, this was the original intent when the term was coined. This would then include former French colonies such as Quebec in Canada, Haiti, Martinique and Guadeloupe in the Caribbean, and French Guiana in South America.
- The former Dutch colonies Suriname, the Netherlands Antilles, and Aruba are not usually considered part of Latin America, even though in the latter two, the predominantly Iberian-influenced language Papiamento is spoken by the majority of the population.
- But sometimes, particularly in the United States, the term "Latin America" is used to refer to all of the Americas south of the U.S., including also countries such as Belize, Jamaica, Barbados, Trinidad and Tobago, Guyana, and Suriname where non-Romance languages prevail.
- Moreover, the scholarly field of Latin American studies increasingly takes in both the French Caribbean, particularly Haiti, (because of the many cultural and historical ties with Spanish America) and also Chicano and Latino culture in the USA.
- Indeed, in historical terms, Latin America could be defined as all those parts of the Americas that were once part of the Spanish or Portuguese (and arguably also French) Empires. Hence much of the US Southwest

plus Florida (and also French Louisiana) would be covered by this definition.

- Finally, it's worth noting that the distinction between Latin and Anglo America, and more generally the stress on European heritage, passes over the fact that there are many places in the Americas (e.g. highland Peru or Guatemala) where non-European cultures and languages are still important, as well as the influence of African cultures in other areas (e.g. the Caribbean, including parts of Colombia and Venezuela, and coastal Brazil).

## Etymology

Originally a political term, Amerique Latine was coined by French emperor Napoleon III, who cited Amerique Latine and Indochine as goals for expansion during his reign. While the term helped him stake a claim to those territories, it eventually came to embody those parts of the Americas that speak Romance languages initially brought by settlers from Spain, Portugal and, to a minor extent, France in the fifteenth and sixteenth centuries. An alternate etymology points to Michel Chevalier, who mentioned the term in 1836.

In the United States, the term was not used until the 1890s, and did not become a common descriptor of the region until early in the twentieth century. Before then, Spanish America was more commonly used.

Latin America has come to represent an expression equivalent to Latin Europe and implies a sense of supra-nationality greater than those implied by notions of statehood or nationhood. This supranational identity is expressed through common initiatives and organizations, like the Union of South American Nations. It is important to observe that the terms Latin American, Latin, Latino, and Hispanic differ from each other.

Many people in Latin America do not speak Latin-derived languages, but native ones or languages brought over by immigration. There is also the blend of Latin-derived cultures with indigenous and African ones resulting in a differentiation in relation to the Latin-derived cultures of Europe.

Quebec, other French-speaking areas in Canada and the United States like Acadia, Louisiana, Saint-Pierre and Miquelon, and other places north of Mexico are traditionally excluded from the sociopolitical definition of Latin America, despite having significant or predominant populations that speak a Latin-derived language, due in part to these territories' not existing as sovereign states or being geographically separated from the rest of Latin America. French Guiana, however, is sometimes included, despite being a dependency of France and not an independent country.

As alluded to above, the term Ibero-America is sometimes used to refer to the nations that were formerly colonies of Spain and Portugal, as these two countries are located on the Iberian peninsula. The Organization of Ibero-American States (OEI) takes this definition a step further, by including Spain and Portugal (often termed the Mother Countries of Latin America) among its member states, in addition to their Spanish and Portuguese-speaking former colonies in America.

**History**

The Americas are thought to have been first inhabited by people crossing the Bering Land Bridge, now the Bering strait, from northeast Asia into Alaska more than 10,000 years ago. Over the course of millennia, people spread to all parts of the continents. By the first millennium AD/CE, South America's vast rainforests, mountains, plains and coasts were the home of tens of millions of people. Some groups formed permanent settlements, such as the Chibchas (or "Muiscas" or "Muyscas") and the Tairona groups. The Chibchas of Colombia, the Quechuas of Peru and the Aymaras of Bolivia were the three Indian groups that settled most permanently.

The region was home to many indigenous peoples and advanced civilizations, including the Aztecs, Toltecs, Caribs, Tupi, Maya, and Inca. The golden age of the Maya began about 250, with the last two great civilizations, the Aztecs and Incas, emerging into prominence later on in the early fourteenth century and mid-fifteenth centuries, respectively.

With the arrival of the Europeans following Christopher Columbus's voyages, the indigenous elites, such as the Incans and Aztecs, lost power to the Europeans. Hernán Cortés destroyed the Aztec elite's power with the help of local groups who disliked the Aztec elite, and Francisco Pizarro eliminated the Incan rule in Western South America. European powers, most notably Spain and Portugal, colonized the region, which along with the rest of the uncolonized world was divided into areas of Spanish and Portuguese control by the Line of Demarcation in 1493, which gave Spain all areas to the west, and Portugal all areas to the east (the Portuguese lands in America subsequently becoming Brazil). By the end of the sixteenth century, Europeans occupied large areas of Central and South America, extending all the way into the present southern United States. European culture and government was imposed, with the Roman Catholic Church becoming a major economic and political power, as well as the official religion of the region.

Diseases brought by the Europeans, such as smallpox and measles, wiped out a large proportion of the indigenous population, with epidemics of diseases reducing them sharply from their prior populations. Historians cannot determine the number of natives who died due to European diseases, but some put the figures as high as 85% and as low as 20%. Due to the lack of written records, specific numbers are hard to verify. Many of the survivors were forced to work in European plantations and mines. Interracial marriage between the indigenous peoples and the European colonists was very common, and, by the end of the colonial period, people of mixed ancestry (mestizos) formed majorities in several colonies.

By the end of the eighteenth century, Spanish and Portuguese power waned as other European powers took their place, notably Britain and France. Resentment grew over the restrictions imposed by the Spanish government, as well as the dominance of native Spaniards (Iberian-born peninsulares) over the major institutions and the majority population, including the Spanish descended Creoles (criollos). Napoleon's invasion of Spain in 1808 marked the turning point, compelling Creole elites to form juntas that advocated independence. Also, the newly independent Haiti, the second oldest nation in the New World after the United States, further fueled the independence movement by inspiring the leaders of the movement, such as Simón Bolívar and José de San Martin, and by providing them with considerable munitions and troops. Fighting soon broke out between the Juntas and the Spanish authorities, with initial Creole victories, such as Father Miguel Hidalgo's in Mexico and Francisco de Miranda's in Venezuela, crushed by Spanish troops. Under the leadership of Simón Bolívar, José de San Martin and other Libertadores, the independence movement regained strength, and by 1825, all of Spanish Latin America, except for Puerto Rico and Cuba, gained independence from Spain. Brazil achieved independence with a constitutional monarchy established in 1822. During the same year in Mexico, a military officer, Agustín de Iturbide, led conservatives who created a constitutional monarchy, with Iturbide as emperor (shortly followed by a republic).

## Political Divisions

Latin America is politically divided into the following countries and territories:

- Independent Countries
- French dependencies
- Netherlands dependencies
- United States dependencies
- Argentina
- Bolivia
- Brazil
- Chile
- Colombia
- Costa Rica
- Cuba
- Dominican Rep.
- Ecuador
- El Salvador
- Guatemala
- Haiti
- Honduras
- Mexico
- Nicaragua
- Panama
- Paraguay
- Peru
- Uruguay
- Venezuela
- French Guiana
- Guadeloupe
- Martinique
- Aruba
- Bonaire
- Curaçao
- U.S. Virgin Islands
- Puerto Rico.

In addition, some might add Belize, the Falkland Islands, Guyana, Trinidad and Tobago, and Suriname to this list, but they are not culturally or linguistically Latin American - although much of Belize's population is. They maintain economic ties with nearby countries, and are grouped by the United Nations in the predominantly Latin American region (South). However, all except Suriname are also the objects of long-standing territorial claims by their Latin American neighbors.

## Population

In Peru, Bolivia and Guatemala the Amerindians make up the majority of the population, although in all Latin American countries there are people of mixed Indian descent

Many Latin Americans are of European descent, mainly of Spanish, Portuguese or Italian descent. Predominantly White countries and regions include Southeastern and Southern Brazil, Argentina and Uruguay.

The population of Latin America is an amalgam of ancestries and ethnic groups. The composition varies from country to country. Some have a predominance of a mixed population, some have a high percentage of people of Amerindian origin, some are dominated by inhabitants of European origin and some populations are primarily of African origin. Most or all Latin American countries have Asian minorities.

## Demographics

In three countries the Amerindians make up the largest segment of the population: in Guatemala and Bolivia they represent a majority of over 50%, and in Peru they constitute a plurality of just under 50%. In the rest of the Region, most people with a Native American lineage are admixed with one or more other ethnic lineages.

Since the sixteenth century a large number of Iberian colonists left for Latin America: the Portuguese to Brazil and the Spaniards to the rest of the region. Intensive mixing between the Europeans and the Amerindians occurred and their descendants, known as mestizos, make up the majority of the population in half of the Latin American countries: Colombia,

Ecuador, El Salvador, Honduras, Nicaragua, Panama, and Venezuela. There's genetic evidence that Puerto Rico may have a mestizo majority as well.

Starting in the late sixteenth century, a large number of African slaves were brought to Latin America, the majority of whom were sent to the Caribbean and Brazil. Nowadays, African descendants make up the majority of the population in most Caribbean countries. Mixing between Africans and Amerindians also occurred and their descendants are known as Zambos, found primarily in Venezuela and Colombia. Many of the African slaves in Latin America mixed with the Europeans, and their descendants, known as Mulattoes, make up the majority of the population in some countries such as the Dominican Republic and Cuba, and a large proportion of the populations of Brazil, Colombia, Venezuela, and Belize. Many Latin American countries also have a substantial "tri-ethnic" population, their ancestry being a mix of European, Amerindian, and African, most notably in Dominican Republic, Colombia, Puerto Rico Venezuela, and Brazil.

Millions of European immigrants arrived in Latin America in the late nineteenth and early twentieth centuries, with most of them settling in Argentina, Brazil, Chile and Uruguay. The top five groups of European immigrants were Spaniards, Portuguese, Italians, Germans and Poles. The descendants of these immigrants and the descendants of Spanish and Portuguese colonial settlers together compose some 90% of the current white Latin American population. Some of the other groups are Russians, Welsh, Ukrainians, French, Croatians and European Jews. More than two thirds of Latin America's entire white population resides in a continuous area of South America that consists of Argentina, southern Brazil and Uruguay. (See Immigration to Argentina and Immigration to Brazil.)

In this same period, many immigrants came from the Middle-East and Asia, including Indians, Lebanese, Syrians, and, more recently, Koreans, Vietnamese, Chinese, and Japanese (mainly to Brazil). In the late nineteenth century, a small wave of Americans, mostly from the former Confederate States of the Southern U.S., settled in Brazil, and fewer across Latin America.

## Racial Origins

These figures include 19 of the 20 Latin American nations. Venezuela is not included as it does not include race on its census.

Total Population 522.8 million. Racial groups: 217 million White (33.3% of the total population), 133.8 million Mestizo (25.6%), 90.3 million Mulatto (17.3%), 60.8 million Amerindian or Native Peoples (11.6%), 31.5 million White/Mestizo (6%; a few countries count Whites and Mestizos together), 24.8 million Black (4.7%), 1.4 million Asian (0.3%; the correct figure may be at least 4.4 million, or.8%), 6.2 million Other and Unknown (1.2%). (Note: Venezuela's population is 26,749,000. Applying to this the country's 1998 ratios [mestizo 67%, white 21%, black 10%, Amerindian 2%] yields, for the entire region: Population 549,552,000; White 33.7%, Mestizo 27.6%, Mulatto 16.4%, Amerindian or Native Peoples 11.2%, White/Mestizo 5.7%, Black 5%, Asian 0.3%, Other and Unknown 1.1%). As these numbers show, although almost every Latin American country has a majority population, that is not the case for the region as a whole. They also show that more than 80% of Latin America's population ranges from those with an amount of fully European origin to those of significant European admixture.

## Language

### *Portugal*

Spanish is the predominant language in the majority of the countries. Portuguese is spoken primarily in Brazil, where it is both the official and the national language. French is also spoken in smaller countries, in the Caribbean, and in French Guiana. Dutch is the official language on various Caribbean islands and in Suriname on the continent; however, as Dutch is a Germanic language, these territories are generally not considered part of Latin America.

Several nations, especially in the Caribbean, have their own Creole languages such as Haiti in which their Creole is a mixture of French, and African tongues along with Spanish and Indian influences to a lesser extent. The Creole languages of Latin America derived from European languages and various

African tongues. Native American languages are spoken in many Latin American nations, mainly Peru, Guatemala, Bolivia, Paraguay, and to a lesser degree in Mexico, Ecuador and Chile. Note that the lesser degree of indigenous speakers in Mexico is proportional to that country's population.

In real numbers, however, Mexico harbors the largest population of indigenous speaker of any country in the Americas, surpassing Amerindian majority countries of Guatemala, Bolivia and the Amerindian plurality country of Peru. The population of speakers of indigenous languages in other countries is tiny or non-existent.

In Peru, Quechua holds official language status, alongside Spanish and any other indigenous language in the areas where they predominate. In Ecuador, while holding no official status, the closely-related Quichua is a recognized language of the indigenous people under the country's constitution; however, it is only spoken by a few groups in the country's highlands. In Bolivia, Aymara, Quechua and Guarani hold official status alongside Spanish.

Guarani is, along with Spanish, the official language of Paraguay, and is spoken by a majority of the population who are for the most part mestizos bilingual in Spanish. In Nicaragua, Spanish is the official language, but on the Caribbean coast English and indigenous languages such as Miskito, Sumo, and Rama (among others) hold official status.

Colombia, recognizes all indigenous languages spoken within its territory as official, though fewer than 1% of its population are native speakers. Nahuatl is only one of the 62 native languages spoken by indigenous people in Mexico, which are officially recognized by the government as "national languages", along with Spanish.

European languages, other than Spanish and Portuguese, that are spoken include; Italian in Brazil, Argentina, Uruguay and to a lesser extent Venezuela; German in southern Brazil, Argentina, and two German-speaking villages, one in southern Chile and another in northern Venezuela; Welsh in southern Argentina.

## Religion

Although most of Latin America is Roman Catholic, membership in the Roman Catholic Church in Latin America is declining while membership in Protestant churches are increasing. Brazil, still has an active quasi-socialist Roman Catholic movement known as Liberation Theology. Practitioners of the Buddhist, Jewish, Islamic, Hindu, Bahá'í, and indigenous denominations and religions exist. Various Afro-Latin American traditions, such as Santería, and Macumba, a tribal- voodoo religion are also practiced.

## Economy

According to ECLAC, an economic growth rate of 5.3% is estimated for 2006, equivalent to a per capita increase of 3.8%. This marks the fourth consecutive year of economic growth, and the third consecutive year of rates exceeding 4%, after an average annual growth rate of only 2.2% between 1980 and 2002. A breakdown of the annual rates of GDP growth (in US dollars at constant 2000 prices) is transcribed as follows:

| Country | 2004 | 2005 | 2006[a] | 2007[b] |
|---|---|---|---|---|
| Latin America | 6.0 | 4.5 | 5.3 | 4.7 |
| Argentina | 9.0 | 9.2 | 8.5 | 7.5 |
| Bolivia | 3.9 | 4.1 | 4.5 | 4.0 |
| Brazil | 5.7 | 2.9 | 3.7 | 4.5 |
| Chile | 6.2 | 6.3 | 4.4 | 6.0 |
| Colombia | 4.9 | 5.2 | 6.0 | 5.0 |
| Costa Rica | 4.1 | 5.9 | 6.8 | 5.0 |
| Cuba[c] | 5.4 | 11.8 | 12.5 | n.a. |
| Dominican Republic | 2.7 | 9.2 | 10.0 | 7.0 |
| Ecuador | 7.9 | 4.7 | 4.9 | 4.0 |
| El Salvador | 1.8 | 2.8 | 3.8 | 4.0 |
| Guatemala | 2.7 | 3.2 | 4.6 | 5.0 |
| Haiti | -3.5 | 1.8 | 2.5 | 3.0 |
| Honduras | 5.0 | 4.1 | 5.6 | 5.0 |
| Mexico | 4.2 | 3.0 | 4.8 | 3.8 |

**Contd...**

| Country | 2004 | 2005 | 2006[(a)] | 2007[(b)] |
|---|---|---|---|---|
| Nicaragua | 5.1 | 4.0 | 3.7 | 4.0 |
| Panama | 7.5 | 6.9 | 7.5 | 7.0 |
| Paraguay | 4.1 | 2.9 | 4.0 | 3.5 |
| Peru | 5.2 | 6.4 | 7.2 | 6.0 |
| Uruguay | 11.8 | 6.6 | 7.3 | 6.0 |
| Venezuela | 17.9 | 9.3 | 10.0 | 7.0 |

(a) Estimate

(b) Projection

(c) Figures provided by the National Statistics Office of Cuba, under evaluation by ECLAC

**Source:** ECLAC

Growth continues to fall short of other developing regions, however. With the international environment remaining favorable, the volume of goods and services exports was up by 8.4% for the region as a whole and the main export prices rose, which translated into a terms-of-trade improvement equivalent to over 7%.

According to the World Bank in 2006 Latin America had higher export revenues and volumes given the record-high commodity world prices than the previous year. Total GDP growth averaged 4.4% in 2005 and it is expected to grow 4.6% in 2006.

The biggest exporter in the region is Mexico; in 2005 Mexico alone exported 213.7 billion USD, roughly equivalent to the exports of all members of Mercosur combined (including Venezuela), which totaled 214.5 billion USD.

In the same year, Mexico also had the largest Gross National Income (GNI) of 753 billion USD and the largest income per capita in the region, 7,310 USD. However, adjusted to purchasing power parity (PPP) instead of using nominal exchange rates, Brazil's Gross Domestic Product (GDP) was the largest in the region (1.701 trillion USD) though Argentina's income per capita in PPP was the largest (13,920 USD).

As a result of these income gains, and of increased remittances from abroad, growth in national income (7.2%) again exceeded GDP expansion. In addition, other factors, such as growing investor and consumer confidence after several years of sustained growth, real interest rates that remained relatively low despite recent hikes in many countries, a stronger boost to public spending, an expansion in total wages driven by rising employment and a modest upturn in real wages, have helped to make domestic demand into an additional engine for growth. In fact, domestic demand rose by 7.0%, with gross domestic investment up by 10.5% and consumption by 6.0%.

Public spending rose in several countries as a result of larger investments in physical and social infrastructure and higher current spending. But since fiscal revenues climbed even more steeply, the prevailing picture shows central governments with higher primary surpluses (up from 1.7% to 2.2% of GDP as a simple average of 19 countries) and narrower overall deficits (from 1.1% to 0.3% of GDP). Alert to changes in international interest rates and to the effects of surging domestic demand and rising fuel prices, many countries' monetary authorities raised benchmark rates, especially in the first half of the year. In most cases, this did not slow economic activity, given the abundant liquidity. Nevertheless, inflation decreased in most of the countries and, in weighted terms, it came down from 6.1% in 2005 to 4.8% in 2006. Many countries had to deal with downward pressure on the exchange rate because of large inflows of foreign currency generated by stronger export prices or remittances. They took different steps to contain the effects of these inflows but, overall, most local currencies appreciated slightly (3.5% on average).

Fueled by sustained economic growth, job creation continued, especially in waged employment. A half percentage point increase in the employment rate was partially offset by a rise in labor market participation. As a result, open unemployment continued the downward trend begun in 2004, albeit more slowly, with a drop of 0.4 percentage points taking the rate to 8.7%. In contrast to the pattern of the last few years, real wages also benefited from increased demand for labor in 2006 and formal sector wages rose by some 3% as a regional average.

The value of the region's merchandise exports rose by 21% and its imports by 20%. Together with higher transfers (over US$ 9 billion net), this improvement of the balance of trade in goods was more than enough to offset the widening deficit on the factor and non-factor services accounts. Hence, the balance-of-payments current account surplus increased from 1.5% of GDP in 2005 to 1.8% in 2006. The capital and financial account surplus was smaller than the previous year, at US$ 230 million. This result reflected external debt-reduction policies, together with the development of domestic financial markets and the accumulation of assets abroad. It also reflected a sharp fall in net foreign investment, which owed much to the Brazilian acquisition of a Canadian firm, while capital flows into the region in the form of foreign direct investment were down slightly in comparison to 2005. The average region-wide performance masks large differences between and within countries. In particular, the international environment has affected exporters of high-demand natural resources, especially in South America (and petroleum-exporting countries in other subregions), in a very different way to the other Latin American and Caribbean countries.

In the light of the risks for the region's future economic development, particularly the risk of a hard or soft landing in the global economy, many countries in the region have taken steps to reduce their vulnerability. Such measures include adopting more flexible exchange-rate regimes, paying down foreign debt, restructuring debt in favour of longer profiles and fixed rates, building up international reserves, strengthening fiscal accounts and reducing the dollarization of their financial systems. Nevertheless, there is no doubt that a global economic slowdown would seriously affect the region's growth and the well-being of its population.

Economic expansion is expected to slow slightly in 2007, with the regional GDP growth rate projected at around 4.7%. If these projections are borne out, the region's per capita output will show a cumulative gain of some 15%, or 2.8% per year, in the period 2003-2007.

Inequality and poverty continue to be the region's main challenges; according to the ECLAC Latin America is the most unequal region in the world. Moreover, according to the World Bank, nearly 25% of the population lives on less than 2 USD a day. The countries with the highest inequality in the region (as measured with the Gini index in the UN Development Report) in 2006 were Bolivia (60.1) Colombia (58.6), Paraguay (57.8) and Chile (57.1), while the countries with the lowest inequality in the region were Nicaragua (43.1), Ecuador (43.7), Venezuela (44.1) and Uruguay (44.9). One aspect of inequality and poverty in Latin America is unequal access to basic infrastructure. For example, access to water and sanitation in Latin America and the quality of these services remain low.

The major trade blocs or agreements in the region Mercosur and the Andean Community of Nations (CAN). Minor blocs or trade agreements are the G3 and the Central American Free Trade Agreement (CAFTA). However, major reconfigurations are taking place along opposing approaches to integration and trade; Venezuela has officially withdrawn from both the CAN and G3 and it has been formally admitted into the Mercosur (pending ratification from the Brazilian and Paraguayan legislatures). The president-elect of Ecuador has manifested his intentions of following the same path. This bloc nominally opposes any Free Trade Agreement (FTA) with the United States, although Uruguay has manifested its intention otherwise. On the other hand, Mexico is a member of the NAFTA, Chile has signed a FTA with the United States, and Colombia's and Peru's legislatures have approved a FTA with the United States and are awaiting its ratification by the US Senate.

The following table lists (in alphabetical order) all the countries in Latin America indicating Gross National Income/ Gross National Product (GNI/GNP), per capita income in nominal terms and adjusted to purchasing power parity (PPP), Gross Domestic Product in PPP, a measurement of inequality through the Gini index (the higher the index the more unequal the income distribution is), and the Human Development Index (HDI). GNI statistics come from the World Bank with data as of 2005. GDP statistics come from the International Monetary Fund with data as of 2006. Gini index and HDI come from the UN Development Program. Green cells indicate the 1st rank in each category, while yellow indicate the last rank.

| Country | GNI[8] | GNI per capita[9] | GNI (PPP) per capita[9] | GDP (PPP)[13] | Income equality[12] | HDI |
|---|---|---|---|---|---|---|
| | *million USD* | *USD* | *USD* | *million USD* | Gini index | |
| Argentina | 173,020 | 4,470 | **13,920** | 621,070 | 52.8 | **0.863** |
| Bolivia | 9,271 | 1,010 | 2,740 | 27,957 | 60.1 | 0.692 |
| Brazil | 644,133 | 3,460 | 8,230 | **1,701,183** | 58 | 0.792 |
| Chile | 145,205 | **8,864** | 12,983 | 212,671 | 57.1 | 0.859 |
| Colombia | 104,520 | 2,290 | 7,420 | 378,435 | 58.6 | 0.790 |
| Costa Rica | 19,867 | 4,590 | 9,680 | 51,089 | 49.9 | 0.841 |
| Cuba[14] | n.a. | n.a. | n.a. | 44,540 | n.a. | 0.826 |
| Dominican Republic | 21,080 | 2,370 | 7,150 | 76,573 | 51.7 | 0.751 |
| Ecuador | 34,759 | 2,630 | 4,070 | 64,671 | 43.7 | 0.765 |
| El Salvador | 16,832 | 2,450 | 5,120 | 38,617 | 52.4 | 0.729 |
| Guatemala | 30,259 | 2,400 | 4,410 | 60,766 | 55.1 | 0.673 |
| Haiti | 3,876 | 450 | 1,840 | 15,554 | 59.2 | 0.482 |
| Honduras | 8,586 | 1,190 | 2,900 | 23,183 | 53.8 | 0.683 |
| Mexico | **753,394** | 7,310 | 10,030 | 1,171,506 | 49.5 | 0.821 |
| Nicaragua | 4,968 | 910 | 3,650 | 22,723 | **43.1** | 0.698 |
| Panama | 14,951 | 4,630 | 7,310 | 27,551 | 56.4 | 0.809 |
| Paraguay | 7,854 | 1,180 | 4,970 | 31,213 | 57.8 | 0.757 |
| Peru | 73,045 | 2,610 | 5,830 | 185,591 | 54.6 | 0.767 |
| Uruguay | 15,096 | 4,360 | 9,810 | 37,267 | 44.9 | 0.851 |
| Venezuela | 127,799 | 4,810 | 6,440 | 193,196 | 44.1 | 0.784 |

## Culture

The rich mosaic of Latin American cultural expressions is the product of many diverse influences:

- Native cultures of the peoples that inhabited the continents prior to the arrival of the Europeans.
- European cultures, brought mainly by the Spanish, the Portuguese and the French. This can be seen in any expression of the region's rich artistic traditions, including painting, literature and music, and in the realms of science and politics. The most enduring European colonial influence was language. Italian and British influençe has been important as well.
- African cultures, whose presence derives from a long history of New World slavery. Peoples of African descent have influenced the ethno-scapes of Latin America and the Caribbean. This is manifest for instance in dance and religion, especially in countries such as Brazil, Venezuela, Colombia, and Cuba.
- The United States, particularly via mass culture such as cinema and TV.

## Literature

Pre-Columbian cultures were primarily oral, though the Aztecs and Mayans, for instance, produced elaborate codices. Oral accounts of mythological and religious beliefs were also sometimes recorded after the arrival of European colonizers, as was the case with the Popol Vuh. Moreover, a tradition of oral narrative survives to this day, for instance among the Quechua-speaking population of Peru and the Quiché of Guatemala.

From the very moment of Europe's "discovery" of the continent, early explorers and conquistadores produced written accounts and crónicas of their experience--such as Columbus's letters or Bernal Díaz del Castillo's description of the conquest of Mexico. During the colonial period, written culture was often in the hands of the church, within which context Sor Juana Inés de la Cruz wrote memorable poetry and philosophical essays. Towards the end of the 18th Century and the beginning of the 19th, a distinctive criollo literary tradition emerged, including the first novels such as Lizardi's El Periquillo Sarmiento (1816).

The 19th Century was a period of "foundational fictions" (in critic Doris Sommer's words), novels in the Romantic or Naturalist traditions that attempted to establish a sense of national identity, and which often focussed on the indigenous question or the dichotomy of "civilization or barbarism" (for which see, say, Domingo Sarmiento's Facundo (1845), Juan León Mera's Cumandá (1879), or Euclides da Cunha's Os Sertões (1902)).

At the turn of the 20th century, modernismo emerged, a poetic movement whose founding text was Rubén Darío's Azul (1888). This was the first Latin American literary movement to influence literary culture outside of the region, and was also the first truly Latin American literature, in that national differences were no longer so much at issue. José Martí, for instance, though a Cuban patriot, also lived in Mexico and the USA and wrote for journals in Argentina and elsewhere.

However, what really put Latin American literature on the global map was no doubt the literary boom of the 1960s and

1970s, distinguished by daring and experimental novels (such as Julio Cortázar's Rayuela (1963)) that were frequently published in Spain and quickly translated into English. The Boom's defining novel was Gabriel García Márquez's Cien años de soledad (1967), which led to the association of Latin American literature with magic realism, though other important writers of the period such as Mario Vargas Llosa and Carlos Fuentes do not fit so easily within this framework. Arguably, the Boom's culmination was Augusto Roa Bastos's monumental Yo, el supremo (1974). In the wake of the Boom, influential precursors such as Juan Rulfo, Alejo Carpentier, and above all Jorge Luis Borges were also rediscovered.

Contemporary literature in the region is vibrant and varied, ranging from the best-selling Paulo Coelho and Isabel Allende to the more avant-garde and critically acclaimed work of writers such as Diamela Eltit, Ricardo Piglia, or Roberto Bolaño. There has also been considerable attention paid to the genre of testimonio, texts produced in collaboration with subaltern subjects such as Rigoberta Menchú. Finally, a new breed of chroniclers is represented by the more journalistic Carlos Monsiváis and Pedro Lemebel.

The region boasts five Nobel Prizewinners: in addition to the Colombian García Márquez (1982), also the Chilean poet Gabriela Mistral (1945), the Guatemalan novelist Miguel Ángel Asturias (1967), the Chilean poet Pablo Neruda (1971), and the Mexican poet and essayist Octavio Paz (1990).

## INDOSPHERE

Indosphere is a term, defined as "a socio-political sphere subsuming those countries, cultures, and languages that have historically come under influence from the politics, culture, religion, and languages of India." Beyond the Indian subcontinent, mainland Southeast Asia was the other recipient center of Indian-influenced cultures, literature, philosophy, political systems, architecture, music, and religion (Hinduism and Theravada Buddhism). The latter region includes notably: Myanmar (Burma), Laos, Thailand, and Cambodia, although Indonesia and Malaysia too absorbed much Indian influence before the coming of Islam to Southeast Asia. The cultures of

India are also imprinted through the Indian diaspora to other countries in the world, and may be particuarly influential in states like Fiji, Mauritius and Guyana where they are a substantially large minority.

## SINOSPHERE

Sinosphere, also known as Chinese world or Chinese character cultural sphere (Simplified Chinese: Traditional Chinese: Pinyin: Hanzi wenhua quan; Japanese: kanji bunkaken; Korean: hanja munhwagweon) is a grouping of countries and regions that are currently inhabited with a significant number of people of Chinese descent or historically under Chinese cultural influence. James C. Bennett, founder of The Anglosphere Institute, sees it as a network commonwealth between Chinese people around the world. Bennett envisages the Sinosphere as consisting of Greater China, and to some extent, its overseas Chinese population in countries like Singapore. One of the main unifying links is based on the Chinese language.

In East Asian commentator circles, the term Chinese cultural sphere or Chinese character cultural sphere is used interchangeably for Sinosphere but covering a broader definition. Chinese cultural sphere denotes a grouping of countries, regions, and people with Chinese cultural legacies. This includes the Sinosphere under the Bennett definition plus countries that have extensive Chinese cultural heritage or are with significant Chinese populations in modern times, including Japan, Korea (North and South), Singapore, and Vietnam. In French, the term le monde chinois (the Chinese world) is used for this concept.

### Modern Origins: 1990s

The concept of Sinosphere, as a network commonwealth, predated the popularization of the modern idea of Anglosphere (Pinyin: Yingyu wenhua quan, "English language cultural sphere") in the English-speaking world, and developed largely independent of the Anglosphere. In the early post-Cold War period of the 1990s, economic reforms in the People's Republic of China, coupled with its recognition as a potent rival government of the Republic of China (Taiwan), increased economic and cultural exchanges between China and overseas Chinese itself, led to emergence of the concept of a network of Chinese people that transcend traditional national borders, political differences, and geographical distances.

Later on, this definition was broadened to include East Asian countries that had historical heritage influenced by China, countries such as Japan, North Korea, South Korea, and Vietnam have increased their economic and cultural contacts with the Chinese-speaking communities in both breadth and scope.

Sometimes in East Asia the term Sinosphere is used to imply the concept of East Asian integration.

## Defining Characteristics

Bennett considers the Sinosphere is unified by first language ability in Chinese. Asian commentators define the unifying factor as influence of traditional Chinese cultural beliefs, marked by Confucianism, Daoism, Mahayana Buddhism, Zen/Chan Buddhism, and the use of Chinese characters as a major part of writing system (Hanzi in Chinese, kanji in Japanese, hanja in Korean, and Hánts in Vietnamese).

## Current Developments

The concept of Sinosphere seemed to undergo a setback with the Asian financial crisis and the advent of the dotcom economies in 1997 and 1998. However, with China's membership in the WTO and continuing economic development there are some repopularizations of the use of the term Sinosphere.

The development of the Anglosphere provides an interesting contrast. The idea of a network commonwealth is common to both Sinosphere and Anglosphere, but the two visions were developed independently from each other. As of the first decade of the 21st century, the concept of the Anglosphere remains at large invisible among Asian commentators supportive of the Sinosphere. Among the few who have heard the concept, the common response is either derision or fear. Those who regard the Anglosphere with scorn take the route that regionalist consolidations will triumph over cultural affinities that are separated by geographical distances, and the Sinosphere is more consolidated on a geographical sense than the more dispersed Anglosphere, and also because they regard China's economic might will overtake the Anglosphere in the not too distant future. Detractors who see the Anglosphere as a threat to the Sinosphere regard the Anglosphere as a concept of Anglo-Saxon imperialism and hegemony, and translate the term into Chinese as (literally "Anglo-Saxon Co-Prosperity Sphere") in an attempt to evoke the memory over the historical Japanese concept of a Greater East Asian Co-Prosperity Sphere during the World War II. Some other commentators point out India is an Anglosphere member and it has the potential to overtake China in economic developments. Some also see the Anglosphere's flexible nature and civil society base as points of strengths that the Sinosphere lacks and which will guarantee it will pose a serious competition against the Sinosphere.

Currently Singapore, Hong Kong, and Japan are contested by both Anglosphere and Sinosphere proponents as under each respective sphere of influence. Singapore has a 76% Chinese majority, but its governmental, legal and business practice are more akin to English-speaking countries courtesy of its British colonial past. Hong Kong's position is similar to Singapore but its population is 98% Chinese and in Hong Kong, Chinese rather than English is commonly used as the daily communication medium. Japan has had ancient Chinese influence ever since the Taika Reform period; however, the Anglosphere has displaced China in influence from the time of Commodore Matthew Perry's visit in 1853. Post-World War II, Japanese political and military interests are more often aligned with the United States than with China.

# 6

# GEOGRAPHY OF ARAB WORLD

**The Arab World Arabic:** (al-`alam al-`arabi) stretches from the Atlantic Ocean in the west to the Persian Gulf in the east, and from the Mediterranean Sea in the north to Central Africa and the Indian Ocean in the south. Consisting of 23 countries with a combined population of some 325 million people, and spanning two continents, it is the largest geo-cultural unit in the world after Russia.

## Language, Politics, and Religion

The Arabic language forms a unifying feature of the Arab World. Though different areas use local dialects of Arabic, all share in the use of the standard classical language (see diglossia). This contrasts with the situation in the wider Islamic world, where Arabic retains its cultural prestige primarily as the language of religion and of theological scholarship, but the populace generally do not speak Arabic languages. The linguistic and political denotation inherent in the term "Arab" is generally dominant over genealogical considerations. Thus, individuals with little or no direct ancestry from the Arabian peninsula (e.g., black Africans, Berbers) could be considered Arabs by virtue of their mother tongue (see Who is an Arab?).

The Arab League, a political organization intended to encompass the Arab World, defines as Arab,

> *"a person whose language is Arabic, who lives in an Arabic speaking country, who is in sympathy with the aspirations of the Arabic speaking peoples."*

The Arab League's main goal is to unify politically the Arab populations so defined. Its permanent headquarters are located in Cairo. However, it was moved temporarily to Tunis during the 1980s, after Egypt was expelled due to the Camp David Accords (1978).

The majority of people in the Arab World adhere to Islam and the religion has official status in most countries. Shariah law exists partially in the legal system in some countries, especially in the Arabian peninsula, while others are secular. The majority of the Arab countries adhere to Sunni Islam. Iraq, however, is a Shia majority country (65%), while Lebanon, Yemen, Kuwait, and Bahrain have large Shia minorities. In Saudi Arabia, the eastern province Al-Hasa region has Shia minority and the southern province city Najran has Ismalia Shiite minority too. Ibadi Islam is practised in Oman and Ibadis make up 75% population of the country.

There are sizable numbers of Christians, living primarily in Lebanon, Egypt, Palestine, Iraq, Jordan, Sudan and Syria. Formerly, there were significant minorities of Arab Jews throughout the Arab World; however, the establishment of the state of Israel prompted their subsequent mass emigration and expulsion within a few decades. Today tiny communities of Jews remain, ranging anywhere from ten in Bahrain to 7,000 in Morocco and more than 1,000 in Tunisia. Overall, Arabs make up less than one quarter of the world's 1.4 billion Muslims, a group sometimes referred to as the Islamic world.

Some Arab countries have substantial reserves of petroleum. The Gulf is particularly well-furnished: four Gulf states, Saudi Arabia, the UAE, Kuwait, and Qatar, are among the top ten oil or gas exporters worldwide. In addition, Algeria, Libya, Iraq, Bahrain, Morocco (Western Sahara) and Sudan all have smaller but significant reserves. Where present, these have had significant effects on regional politics, often enabling rentier states, leading to economic disparities between oil-rich and oil-poor countries, and, particularly in the more sparsely populated states of the Gulf and Libya, triggering extensive labor immigration.

According to UNESCO, the average rate of adult literacy (ages 15 and older) in this region is 66%, and this is one of the lowest rates in the world. In Mauritania, Morocco, and Yemen, the rate is lower than the average, at barely over 50 %. On the other hand, Kuwait and Palestine record a high adult literacy rate of over 90%. The average rate of adult literacy shows steady improvement, and the absolute number of adult illiterates fell from 64 million to around 58 million between 1990 and 2000-2004. Overall, the gender disparity in adult literacy is high in this region, and of the illiteracy rate, women account for two-thirds, with only 69 literate women for every 100 literate men. The average GPI (Gender Parity Index) for adult literacy is 0.72, and gender disparity can be observed in Egypt, Morocco, and Yemen. Above all, the GPI of Yemen is only 0.46 in a 53% adult literacy rate.

Literacy rate is higher among the youth than adults. Youth literacy rate (ages 15-24) in the Arab region increased from 63.9 to 76.3 % from 1990 to 2002. The average rate of GCC States was 94 %, followed by the Maghreb at 83.2% and the Mashriq at 73.6 %. However, more than one third of youth remain illiterate in the Arab LDCs (Comoros, Djibouti, Mauritania, Somalia, Sudan, and Yemen). In 2004, the regional average of youth literacy is 89.9% for male and 80.1 % for female.

The average population growth rate in Arab countries is 2.3%.

The United Nations have published an Arab human development report in 2002, 2003 and 2004. These reports, written by Arabic researchers, address some sensitive issues in the development of the Arab countries: women empowerment, availability of education and information.

## Non-Arab Peoples in the Arab World

Within the most common definition of the Arab World, the Arab League's, there reside substantial populations who are not Arab either by ethnic or linguistic definition, and who often or generally do not consider themselves Arab as such. Nevertheless, most are as indigenous as the Arabs to the area, and many, if not most, actually resided in the area before the arrival of Arabs with the spread of Islam. Certain populations have expressed resentment towards the term "Arab World," and believe that their national and political rights have been unjustly brushed aside by modern governments' focus on Pan-

Arabism and promoting an Arab identity. In some cases this has led to severe conflicts between the ethnic nationalism of these groups and the Arab nationalism promoted by governments lead by Arab0identifying leaders, which sometimes amounted to denying the existence of or forcibly suppressing non-Arab minorities within their borders.

In North Africa clearly a majority of the population is of Berbers descent (as opposed to current ethnic identification) as the number of Arabs who settled in North Africa was very small (about 200,000). However, a clear majority now self-identifies as Arab. Berber and Arab identity in these countries is generally defined by primary language use rather than ancestry. In Morocco, Berber speakers form over 35% of the total population; in Algeria, they represent about 20% of the population, more than half in the eastern region of Kabylie. In Libya, they form about 4% of the population, mainly near the Tunisian border. There are much smaller isolated Berber communities in Tunisia, Mauritania, and even one oasis in Egypt.

The nomadic Touareg people whose traditional areas straddle the borders of several countries in the Sahara desert, are also of Berber origins. Government worries about ethnic separatism, and condescending attitudes towards the mainly rural Berber-speaking areas, led to the Berber communities being denied full linguistic and cultural rights; in Algeria, for example, Berber chairs at universities were closed, and Berber singers were occasionally banned from singing in their own language, although an official Berber radio station continued to operate throughout. These problems have to some extent been redressed in later years in Morocco and Algeria; both have started teaching Berber languages in schools and universities, and Algeria has amended its constitution to declare Berber a fundamental aspect of Algerian identity (along with Islam and Arabness.) In Libya, however, any suggestion that Berbers might be non-Arab remains taboo.

In the northern regions of Iraq (15-20%) and Syria (5-10%) live the Kurds, a mountain people who speak Kurdish, a language closely related to Farsi, but not directly to Arabic, except insofar as like Farsi, it has absorbed Arabic vocabulary. The nationalist aspiration for self-rule or for a state of Kurdistan has created conflict between Kurdish minorities and their governments.

Somalia is a non-Arab country. Although, Somalia is a member of the Arab League, the predominant language is Somali, and the population is predominantly Somali (85% of the population, with the remaining 15% Somali Bantu, as well as some small Arab-mixed groups) and ethnic Arabs such as the Benadiri Arabs (who have roots from Arabia through Arab traders coming to the Benadir region). For this reason, they are sometimes excluded from the definition of the Arab world. However, Somalia joined the Arab League in 1974. Arabic is the second official language of Somalia and is used for religious purposes. Over 99% of the country's population is Muslim, resulting in close ties with decidedly Arab states.

Djibouti (which is part of Greater Somalia), is approximately 60% Somali and 35% Afar, is in a similar position. Arabic is one of the official languages, 94% of its population is Muslim, and Djibouti has a close proximity on the Red Sea and Arabia, although it has no native Arab population.

Other examples of non-Arab peoples originating in what is often labelled the Arab World include the Turkmen of Iraq, Assyrians and Jews (most of whom fled to Israel after its creation in 1948). Since most Arab League states are products of colonialism, their borders rarely reflect distinct ethnic or geographic boundaries. Thus, many peripheral states of the Arab World have border-straddling minorities of non-Arab peoples. This is the case with Iranians in Iraq (most of whom fled in the Iraq-Iran War) and the non-Arab black (often called black Africans) peoples in Sudan.

Many Gulf Arab countries have sizable (10 - 30%) non-Arab populations, usually of a temporary nature, at least in theory. Iraq, Bahrain, Kuwait, Qatar, United Arab Emirates and Oman has sizeable Farsi speaking minority. The same countries also have Hindi-Urdu speakers and Filipinos as sizable minority. Balochi speakers are a good size minority in Oman. Countries like Bahrain, UAE, Oman and Kuwait have significant non-Muslim / non-Arab minorities (10 - 20%) like Hindus and Christians from India, Pakistan, Bangladesh, Nepal and the Philippines.

Many countries bordering the core Arab world like Chad, Israel, Iran and Mali have sizable Arab minorities.

**States**

While this has often been a matter of dispute, the following entities (18 states and two territories) are considered part of the Arab World as all use Arabic as one of their official state languages, and most of their inhabitants speak one or more of the Arabic languages and dialects (though ethnically quite varied):

- Algeria
- Bahrain
- Egypt
- Iraq
- Jordan
- Kuwait
- Lebanon
- Somalia
- Djibouti
- Libya
- Mauritania
- Morocco
- Oman
- Palestine
- Qatar
- Saudi Arabia
- Sudan
- Syria
- Tunisia
- United Arab Emirates
- Western Sahara (administered by Morocco)

## Yemen

Palestine is not de facto a sovereign country, but is awaiting statehood and its government, the Palestinian Authority is recognised as a legitimate state by over a hundred countries in addition to being a full-fledged member of the Arab League and many other international organizations. Palestine is an observer member of the UN. The territory of Western Sahara is disputed between the kingdom of Morocco and the Polisario Front which declared a government-in-exile, the Sahrawi Arab Democratic Republic which is a member of the African Union. In addition to the countries and territories listed, Djibouti, Somalia and the Comoros are all member states of the Arab League, although their inhabitants are not predominantly Arabic-speaking (though many use and understand Arabic for religious purposes). The predominate language in Somalia and Djibouti is the Somali language, part of the Afro-Asiatic East Cushitic language group. On the other hand, the Maltese language is closely related to Tunisian Arabic, but Malta does not use standard Arabic and its inhabitants do not consider themselves Arabs. Chad, Eritrea, and Israel all recognize standard Arabic as an official language, but none of them are members of the Arab League (furthermore most Arab League members do not maintain diplomatic relations with Israel). Mali and Senegal recognize Hassaniya, the Arabic dialect of their Moorish minorities, as a national language, as does Egypt with respect to Masri, its own variant of Arabic, but grant them no official status.

Different forms of government are represented in the Arab World: Some of the countries are monarchies: Bahrain, Jordan, Kuwait, Morocco, Oman, Qatar, Saudi Arabia and the United Arab Emirates. The other Arab countries are all republics, and their official names may ostensibly indicate that they are democracies. In practice, however, with the exceptions of Lebanon, Mauritania and maybe Yemen, the Arab states are ruled by either an absolute monarch, an unelected president or a single political party.

The borders of the various states were often drawn up by European colonial powers in the 19th and 20th centuries. They are often straight lines drawn on a map with complete disregard to the geographic and demographic characteristics of the land (see below). After World War II, the movement known as Pan-Arabism sought to unite all Arabic-speaking countries into one political entity. Only Syria, Iraq, Egypt, Tunisia, Libya and North Yemen attempted the short-lived unification. Historical divisions, competing local nationalisms, and geographical sprawl were major reasons for the failure of Pan-Arabism. Arab Nationalism was another strong force in the region which peaked during the mid 20th century and was professed by many leaders in Egypt, Algeria, Libya, Syria, and Iraq. Incidentally, Arab nationalism was unpopular in Egypt in the first decades of the 20th century (until Nasser), and was once again relegated by ethnic Egyptian nationalism after Anwar Sadat's assumption of power.

Arab Nationalist leaders included Gamal Abdel Nasser of Egypt, Ahmed Ben Bella of Algeria, Michel Aflaq, Salah al-Din al-Bitar, Zaki al-Arsuzi, Constantin Zureiq, Shukri al-Kuwatli, Hafez al-Assad and Bashar al-Assad of Syria, Saddam Hussein and Ahmad Hassan al-Bakr of Iraq, Moammar al-Qadhafi of Libya, Habib Bourguiba of Tunisia, Mehdi Ben Barka of Morocco, and Shakib Arslan of Lebanon. The various Arab states maintain close ties but national identities have been strengthened by the political realities of the past 60 years (much longer for Egypt), making a single Arab nationalistic state less and less feasible. The upsurge in Islam and the call for a single Islamic state or Caliphate in the Muslim world has grown leaving Arab nationalists behind. Islamic rhetoric as opposed to rhetoric about an 'Arab nation' holds more political currency among Arabs. Arab nationalists once persecuted Islamic & democratic parties, but now pander to them for political survival.

## Modern Boundaries

The modern borders of the Arab World are considered by Arab nationalists as largely artificial creations of European colonial powers during the 18th, 19th and early 20th century. However, some of the larger states--in particular Egypt and perhaps Syria--have historically maintained geographically definable boundaries, on which some of the modern states are roughly based. The 14th century Egyptian historian Al-Maqrizi, for instance, defines Egypt's boundaries as extending from the

Mediterranean in the north to lower Nubia in the south; and between the Red Sea in the east and the oases of the Western/ Libyan desert. The modern borders therefore are more than mere creations of European colonial powers, and are at least in part based on historically definable entities which are in turn based on certain cultural and ethnic identifications.

At other times, Kings, 'emirs' or 'sheiks' were placed as semi-autonomous rulers over the newly created nation states, usually chosen by the same colonial powers that for some drew the new borders, for services rendered to colonial entities like the British Empire e.g. Sherif Hussein ibn Ali. Many African States did not attain independence until the 1960s from France after bloody insurgencies for their freedom. These struggles were settled by the colonial powers approving the form of independence given, so as a consequence almost all of these borders have remained. Some of these borders were agreed upon without consultation of those individuals that had served the colonial interests of Britain or France. One such agreement solely between Britain and France (to the exclusion of Sherif Hussein ibn Ali), signed in total secrecy until Lenin released the full text, was the Sykes-Picot Agreement. Another influential document written without the consensus of the local population was the Balfour Declaration.

As former director of the Israeli intelligence agency Mossad, Efraim Halevy, now a director at the Hebrew University said,

> *The borders, which if you look on the maps of the middle-east are very straight lines, were drawn by British and French draftsmen who sat with maps and drew the lines of the frontiers with rulers. If the ruler for some reason or other moved on the map, because of some person's hand shaking, then the frontier moved (with the hand).*

He went on to give an example,

> *There was a famous story about a British consul, a lady named Gertrude Bell who drew the map between Iraq and Jordan, using transparent paper. She turned to talk to somebody and as she was turning the paper moved and the ruler moved and that added considerable territory to the (new) Jordanians.*

Historian Jim Crow, of Newcastle University, has said: *Without that imperial carve-up, Iraq would not be in the state it is in today...Gertrude Bell was one of two or three Britons who were instrumental in the creation of the Arab states in the Middle East that were favourable to Britain.*

## Modern Economies

The Arab States are mostly, although not exclusively, developing economies and derive their export revenues from oil and gas, or the sale of other raw materials. Recent years have seen significant economic growth in the Arab World, due largely to an increase in oil and gas prices, which tripled between 2001 and 2006, but also due to efforts by some states to diversify their economic base. Industrial production has risen, for example the amount of steel produced tripled between 2004 and 2005 from 8.4 to 19 million tonnes. (Source: Opening speech of Mahmoud Khoudri, Algeria's Industry Minister, at the 37th General Assembly of the Iron & Steel Arab Union, Algiers, May 2006). However even 19 million tons p.a. still only represents 1.7% of global steel production, and remains inferior to the production of countries like Brazil.

The main economic organisations in the Arab World are the Gulf Cooperation Council (GCC), comprising the Gulf states, and the Union of the Arab Maghreb (UMA), made up of North African States. The GCC has achieved some success in financial and monetary terms, including plans to establish a common currency in the Persian Gulf region. Since its foundation in 1989, the UMA's most significant accomplishment has been the establishment of a 7000 km highway crossing North Africa from Mauritania to Libya's border with Egypt. The central stretch of the highway, expected to be completed in 2010, will cross Morocco, Algeria and Tunisia. In recent years a new term has been coined to define a greater economic region: the MENA region (standing for Middle East and North Africa) is becoming increasingly popular, especially with support from the current US administration.

Saudi Arabia remains the top Arab economy in terms of total GDP. It is Asia's eleventh largest economy, followed by

Egypt and Algeria, which were also the second and third largest economies in Africa (after South Africa), in 2006. In terms of GDP per capita, Qatar is the richest developing country in the world. (Source: CIA World Factbook, GDP by country classification)

## Geography

The Arab world stretches across more than 12.9 million square kilometers (5 million square miles) of North Africa and the part of Western Asia called the Middle East. The Asian part of the Arab world (including Arabia proper) is called the Mashreq. The North African part (excluding Egypt and the Sudan) of the Arab World is known as the Maghreb. Its total area is the size of the entire Spanish-speaking Western Hemisphere (also 12.9 million square kilometers), larger than Europe (10.4 million), Canada (10 million), China (9.6 million), the United States (also 9.6 million), Brazil (8.7 million). Only Russia - at seventeen million square kilometers, the largest country in the world - and arguably Anglophone North America (eighteen million square kilometers) are larger geocultural units.

The term "Arab" often connotes the Middle East, but the larger (and more populous) part of the Arab World is North Africa. Its eight million square kilometers include the two largest countries of the African continent, Sudan (2.5 million square kilometers) in the southeast of the region and Algeria (2.4 million) in the center, each about three-quarters the size of India, or about one-and-a-half times the size of Alaska, the largest state in the United States. The largest country in the Arab Middle East is Saudi Arabia (two million square kilometers). At the other extreme, the smallest autonomous mainland Arab country in North Africa and the Middle East is Lebanon (10,452 square kilometres), and the smallest island Arab country is Bahrain (665).

Notably, every Arab country borders a sea or ocean.

## Historical Boundaries

The political borders of the Arab World have wandered, leaving Arab minorities in non-Arab countries of the Sahel and the Horn of Africa as well as in the Middle Eastern countries of Turkey and Iran, and also leaving non-Arab minorities in Arab countries. However, the basic geography of sea, desert, and mountain provide the enduring natural boundaries for this region.

The Arab World straddles two continents, Africa and Asia, and is oriented mainly along an east-west axis, dividing it into African and Asian (Arabian, Middle Eastern) areas.

**Arab Africa**

Arab Africa-or more commonly Arab North Africa, though this is redundant-is roughly a long trapezoid, narrower at the top, that comprises the entire northern third of the continent. It is surrounded by water on three sides (west, north, and east) and desert or desert scrubland on the fourth (south).

In the west, it is bounded by the shores of the Atlantic Ocean. From northeast to southwest, Morocco, Western Sahara (annexed and occupied by Morocco), and Mauritania make up the roughly 2,000 kilometers of Arab Atlantic coastline. The southwestern sweep of the coast is gentle but substantial, such that Mauritania's capital, Nouakchott (18°N, 16°W), is far enough west to share longitude with Iceland (13-22°W). Nouakchott is the westernmost capital of the Arab World and the third-westernmost in Africa, and sits on the Atlantic fringe of the southwestern Sahara. Next south along the coast from Mauritania is Senegal, whose abrupt border belies the gradient in culture from Arab to black African that historically characterizes this part of West Africa.

Arab Africa's boundary to the north is again a continental boundary, the Mediterranean Sea. This boundary begins in the west with the narrow Strait of Gibraltar, the thirteen kilometer wide channel that connects the Mediterranean with the Atlantic to the west, and separates Morocco from Spain to the north. East along the coast from Morocco are Algeria, Tunisia, and Libya, followed by Egypt, which forms the region's (and the continent's) northeastern corner. The coast turns briefly but sharply south at Tunisia, slopes more gently southeastward through the Libyan capital of Tripoli, and bumps north through Libya's second city, Benghazi, before turning straight east again through Egypt's second city, Alexandria, at the mouth of the

Nile. Along with the spine of Italy to its north, Tunisia thus marks the junction of western and eastern Mediterranean, and a cultural transition as well: west of Tunisia begins the region of the Arab world known as the Maghreb.

Historically the 4,000-kilometer Mediterranean boundary has fluttered. Population centers north of it in Europe have invited contact and Arab exploration-mostly friendly, though sometimes not. Islands and peninsulas near the Arab coast have changed hands. The islands of Sicily and Malta lie just a hundred kilometers east of the Tunisian city of Carthage, which has been a point of contact with Europe since its founding in the first millennium B.C.E.; both Sicily and Malta at times have been part of the Arab world. Just across the Strait of Gibraltar from Morocco, regions of the Iberian peninsula were part of the Arab world throughout the Middle Ages, extending the northern boundary at times to the foothills of the Pyrenees and leaving a substantial mark on local and wider European and Western culture.

The northern boundary of the African Arab World has also fluttered briefly in the other direction, first through the Crusades and later through colonization by France, Britain, Spain, and Italy. Another visitor from northern shores, Turkey, controlled the east of the region for centuries, though not as a colonizer. Spain still maintains two small enclaves, Ceuta and Melilla, along the otherwise Moroccan coast. Overall this wave has ebbed, though like the Arab expansion north it has left its mark. The proximity of North Africa to Europe has always encouraged interaction, and this continues with Arab immigration to Europe and European interest in the Arab countries today. However, population centers and the physical fact of the sea keeps this boundary of the Arab world settled on the Mediterranean coastline.

To the east, the Red Sea defines the boundary between Africa and Asia, and thus also between Arab Africa and the Arab Middle East. This sea is a long and narrow waterway with a northwest tilt, stretching 2,300 kilometers from Egypt's Sinai peninsula southeast to the Bab-el-Mandeb strait between Djibouti in Africa and Yemen in Arabia but on average just 150 kilometers wide. Though the sea is navigable along its length, historically much contact between Arab Africa and the Arab Middle East has been either overland across the Sinai or by sea across the Mediterranean or the narrow Bab al Mendeb strait. From northwest to southeast, Egypt, Sudan, and Eritrea form the African coastline, with Djibouti marking Bab al Mendeb's African shore.

Southeast along the coast from Djibouti is Somalia, but the Somali coast soon makes a 90-degree turn and heads northeast, mirroring a bend in the coast of Yemen across the water to the north and defining the south coast of the Gulf of Aden. The Somali coast then takes a hairpin turn back southwest to complete the horn of Africa. For six months of the year the monsoon winds blow from up equatorial Somalia, past Arabia and over the small Yemeni archipelago of Socotra, to rain on India; they then switch directions and blow back. Hence the east- and especially southeast-coast boundary of Arab Africa has historically been a gateway for maritime trade and cultural exchange with both East Africa and the subcontinent. The trade winds also help explain the presence of the Comoros islands, an Arab-African country, off the coast of Mozambique, near Madagascar in the Indian Ocean, the southernmost part of the Arab World.

The southern boundary of Arab North Africa is the stripe of scrubland known as the Sahel, that crosses the continent south of the Sahara, dipping further south in Sudan in the east.

### Arabia and the Arab Middle East

The Asian or Middle Eastern Arab World comprises the Arabian Peninsula, Bilad al-Sham or the Levant, and Iraq, more broadly or narrowly defined. The peninsula is roughly a tilted rectangle that leans back against the slope of northeast Africa, the long axis pointing toward Turkey and Europe.

## EAST ASIA

East Asia is a subregion of Asia that can be defined in either geographical or cultural terms. Geographically, it covers about 12,000,000 km$^2$, or about 28% of the Asian continent and about 15% bigger than the area of Europe. More than 1.5 billion people, about 40% of the population of Asia or a quarter of all the people in the world, live in geographic East Asia, which is about twice the population of Europe. The region is one of the world's most crowded places. The population density of East Asia, 130 per km², is about three times the world average.

Culturally, it embraces those societies that have long been part of the Chinese cultural sphere:

- displaying heavy historical influence from the Classical Chinese language (including the traditional Chinese script),
- Confucianism and Neo-Confucianism
- Mahayana Buddhism,
- and Taoism (Daoism).

This combination of language, political philosophy, and religion (as well as art, architecture, holidays and festivals, etc.) overlaps with the geographical designation of East Asia for the most part, with a few exceptions, such as the overseas Chinese (including those in Singapore, Malaysia, Vietnam, and the West).

East Asia and Eastern Asia (the latter form preferred by the United Nations) are both more modern terms for the traditional European name the Far East, which describes the region's geographical position in relation to Europe rather than its location within Asia. However, in contrast to the United Nations definition, East Asia commonly is used to refer to the eastern part of Asia, as the term implies. What the UN terms 'East Asia' is often referred to as Northeast Asia.

## Other uses of the Term East Asia

All the countries in Eastern Asia: the countries of Northeast Asia, Southeast Asia, and eastern Siberia. The following countries (both states and territories) are commonly seen as located in geographic East Asia:

- People's Republic of China (China)
  * Hong Kong (a special administrative region of the People's Republic of China)
  * Macau (a special administrative region of the People's Republic of China)

- Republic of China (Taiwan)
- Japan
- Democratic People's Republic of Korea (North Korea)
- Republic of Korea (South Korea)
- Mongolia

The following peoples or societies are commonly seen as being encompassed by cultural East Asia:

- Chinese society (including the Chinese-dominated regions of Hong Kong and Macau, Taiwan, and Singapore, with their ethnically Chinese-majority populations)
- Japanese society
- Korean society
- Mongolian society
- Vietnamese society

Some consider the following countries or regions as part of East Asia, while others do not. Disagreements hinge on the difference between the cultural and geographic definitions of the term. Political perspective is also an important factor. In descending order in terms of the frequency with which they are described as East Asian:

- The parts of China that are not historically dominated by Han Chinese: Qinghai, Tibet, Xinjiang (considered either East Asia or Central Asia or South Asian in the case of Tibet -here the primary question is cultural, with geography also at issue)
- Mongolia (considered either East Asia or Central Asia-here culture and/or geography may be at issue)
- Singapore (considered either East Asia or Southeast Asia-here the primary question is geographic)
- Vietnam (considered either East Asia or Southeast Asia-here the primary question is geographic)
- Russian Far East (considered either East Asia or North Asia-here the primary question is political, with culture and geography also at issue)

In infrequent circumstances, the term East Asia is purposefully used to include all countries in Southeast Asia, especially when used in dualism with the term West Asia, the latter of which is then used to include those regions commonly considered West Asia, Central Asia, and Southwest Asia.

**Other Subregions of Asia**

- Southeast Asia
- South Asia
- Central Asia
- Southwest Asia or West Asia (One definition of the Middle East is synonymous with Southwest Asia)
- North Asia (Siberia)
- Northern Eurasia (Extends into part of Europe)
- Central Eurasia (Extends into part of Europe).

# 7

# Geography of Eastern World

The term Eastern world refers very broadly to the various cultures, social structures and philosophical systems of "the East", namely Asia (including China, India, Japan, the Middle East, and surrounding regions).

## CONCEPT OF "THE EAST"

The division between "East" and "West" is a product of European cultural history, and of the distinction between European Christendom and the alien cultures beyond it to the East. Before the discovery of the Americas and the exploration of Sub-Saharan Africa by the Europeans, only North Africa and other Islamic countries to the East were known in detail, though India and China were vaguely known of. The crusades established what became a border between "Eastern" and "Western" peoples. With the European colonization of the Americas the East/West distinction became global. The concept of an Eastern, "Indian" (Indies) or "Oriental" sphere was emphasized by ideas of racial as well as religious and cultural differences. Such distinctions were articulated by Westerners in the scholarly tradition known as Orientalism and Indology. People from the East are known by certain regions in the West as "Oriental". Before the arrival of Islam in India, people from the Indian subcontinent were generally known as Hindus and the subcontinent was known as Hindustan or Bharat. During the Cold War, the term "Eastern world" was sometimes used as an extension of Eastern bloc, connoting the Soviet Union, China and their communist allies, while the term "Western world" often connoted the United States and its NATO allies such as the United Kingdom and France.

The distribution of the two major families of world religion, Dharmic religion and Abrahamic religion, highlights the cultural difference between the far east and the rest of the world.

The concept is often another term for the Far East - a region that bears considerable cultural and religious commonality. Eastern philosophy, art, literature, and other traditions, are often found throughout the region in places of high importance, such as popular culture, architecture and traditional literature. The spread of Buddhism and Hindu Yoga is partly responsible for this.

## Problems of the Concept

Currently terms such as Western, Near East (or Middle East) and Far East are commonly used to distinguish different cultural spheres, based on the standard two-dimensional layout of the world-map, which has the Americas at the far left (West), Europe and Africa in the middle, and Asia to the right (East). This arrangement is arbitrary because the Earth is round, rather than flat (however this model does ensure that land regions are concentrated in the centre without Eurasia being split in two); therefore, by going west, one will eventually arrive in the east, and when one goes east, one will eventually arrive in the west, provided one keeps going long enough. This is true no matter where one is on the globe's surface (except the Poles, where all directions are either north or south). Because of this, the East/West division has been criticized for being Eurocentric, however the notional 'central-point' between East and West would be to the east of Europe itself. Some countries, in particular Russia, do not fit neatly into this opposition.

While Western Orientalist traditions included both Islamic and further Eastern cultures under the generic heading of "the East", the common Abrahamic traditions of Islam and Christianity mean that a case can be made that both Islam and Christianity together form a different cultural sphere from countries further to the East in which the concept of Dharma plays a far more important role than that of an authoritative God. In recent years the concept of "Eastern culture" has increasingly become restricted to East Asian traditions. However, the existence of Islam and Christianity as powerful forces in countries such as Indonesia and the Philippines makes this usage problematic.

## Foundations of Eastern Culture

Eastern culture has developed many themes and traditions. Some important ones are:

- Buddhism - contributed to a concept of self wherein the importance of ego was diminished, and that suffering was regarded as an accepted part of life.
- Confucianism - social hierarchy where position is based upon merit as determined through the imperial examination system. This social hierarchy promoted respect for elders and authority, and valued harmony within family.
- Hinduism - the oldest surviving religion, which contributed the concept of Dharma and served as the origin of the Dharmic religions
- Islam - the majority of the world Muslim population have always lived in Asia, due to the influence of the Islamic civilization
- Martial arts
- Taoism - concepts of good and bad energy, the connectedness of mind and body and nature, etc.
- "Oriental Medicine" - reflexology, acupuncture, herbal medicine, etc.
- Zoroastrianism - the oldest surviving monotheistic religion that was practiced in ancient Persia

## Achievements of Eastern Culture

Eastern culture, especially China, India, and the Middle East, had and still have a major impact on world civilization.

### China

It was China that invented paper, compass, gunpowder and printing (the Four Great Inventions of ancient China) that had profound effects on civilizations worldwide. Among the technological accomplishments of China were early seismological detectors, dry docks, sliding calipers, the double-action piston pump, cast iron, the iron plough, the multi-tube seed drill, the wheelbarrow, the suspension bridge, the parachute, natural gas as fuel, the raised-relief map, the propeller, and the crossbow. Chinese astronomers were also among the first to record observations of a supernova. It was eastern culture that first invented the abacus, planetarium, book, ink, first cannon, bomb (using gunpowder), nest cart, spaghetti, fireworks, brandy, whiskey, cards, Paper money, saddle, toothbrush,. The blue light-emitting diode was invented in Japan. In medicine, the elixir formulation, herbal medicine, craniotomy, and acupuncture are all attributed to the East. It was Easterners who first discovered methamphetamine, adrenaline, sodium glutamate, and Vitamin B1.

**India**

Four of the world's major religions, Hinduism, Buddhism, Jainism and Sikhism originated in India. Hinduism, the successor of the ancient Vedic religion, is considered to be the world's oldest existing religion. Though Buddhism originated in India, it is one of the most practiced religions in East Asia and South East Asia and helped spread Indian philosophical theories like Karma and Dharma to other parts of Asia. Yoga, a family of ancient spiritual practices, originated in India and is one of the six schools of Hindu philosophy. Indian thinkers made great work and effort in mathematics among others and Indian mathematics described and theorized many mathematical concepts and applications. Indians first invented the decimal notation that is 0, 1, 2, 3, 4,...etc.

India is home to some of the greatest and some of the earliest inventions in mathematical concepts, astronomy, physics, medicine, and applications. It is now generally accepted that India was the birth place of several mathematical concepts, including zero, the decimal system, algebra, algorithm, square root and cube root. It was Indians who theorized about gravity, determining sun is a star among others. Aryabhatta is considered to be the first mathematician to use letters of the alphabet to denote unknown quantities and to conclude the pi is an irrational number. Other Indian mathematicians, such as Brahmagupta and Bhaskara too made significant contributions to field of mathematics. Acharya Sushruta, author of Sushruta Samhita, is believed to be the first person to carry out a surgery. Ayurveda, an ancient Indian medical practice, depicts the achievements Indians had made in the field of medical science. The most commonly held view is that Chess originated in India. For instance Aryabha?a is widely regarded as one of the famous and influential thinkers, mathematicians and astronomers who for instance described the solar eclipse, who first described the reason for movement of stars, who first calculated the circumference of the earth with only 0.2% difference from the current prediction.

## Middle East

### *Ancient Near East*

The terms ancient Near East or ancient Orient encompass the early civilizations predating classical antiquity in the region roughly corresponding to that described by the modern term Middle East (Egypt, Iraq, Turkey), during the time roughly spanning the Bronze Age from the rise of Sumer and Gerzeh in the 4th millennium BCE to the expansion of the Persian Empire in the 6th century BCE. As such, it is a term widely employed in the fields of Near Eastern archaeology, ancient History and Egyptology.

The ancient Near East is generally understood as encompassing Mesopotamia (modern Iraq and Syria), Persia (Iran), Armenia, Egypt, the Levant (Israel, Jordan, Lebanon, Syria, Palestinian Authority), and Anatolia (Turkey). Some users of the term would extend its application into the Caucasus region, into modern Afghanistan (see Bactria, Indus Valley Civilization), Minoan and Mycenaean Greece and other peripheral areas. Still, others would exclude Egypt from the ancient Near East as a geographically and culturally distinct area. However, because of Egypt's intimate involvement with the region, especially from the 2nd millennium BCE, this exclusion is rare.

## Family Life

Mesopotamia was a patriarchial society and the men were far more powerful then the women. As for schooling, only royal offspring and sons of the rich and professionals such as scribes, physicians, temple administrators, and so on, went to school. Most boys were taught their father's trade or were apprenticed out to learn a trade. Girls had to stay home with their mothers to learn housekeeping and cooking, and to look after the younger children. Unusual for that time in history, women in Mesopotamia had rights. They could own property and, if they had good reason, get a divorce.

**Agriculture**

Food supply in Mesopotamia was quite rich due to the location of the two rivers from which its name is derived, Tigris and Euphrates. The Tigris and Euphrates River valleys formed the northeastern portion of the Fertile Crescent, which also included the Jordan River valley & that of the Nile. Although land nearer to the rivers was fertile and good for crops, portions of land further from the water were dry and largely uninhabitable.

This is why the development of irrigation was very important for settlers of Mesopotamia. Other Mesopotamian innovations include the control of water by dams and the use of aqueducts. Early settlers of fertile land in Mesopotamia used wooden plows to soften the soil before planting crops such as barley, onions, grapes, turnips, and apples. Mesopotamian settlers were some of the first people to make beer and wine. The unpredictable Mesopotamian weather was often hard on farmers; crops were often ruined so backup sources of food such as cows and lambs were also kept. As a result of the skill involved in farming in the Mesopotamian, farmers did not depend on slaves to complete farm work for them, with some exceptions. There were too many risks involved to make slavery practical (i.e. the escape/mutiny of the slave).

**Government**

**Kings**

The Mesopotamians believed their kings and queens were descended from the city gods, but, unlike the ancient Egyptians, they never believed their kings were actually gods. Most kings named themselves "king of the universe" or "great king". Another common name was "shepherd", as kings had to look after their people. Nebuchadnezzar was the most powerful king in Babylonia. He was thought to be the son of the god Nabu. He married the daughter of Cyaxeres, so the Median and the Babylonian dynasties had a familial connection. Nebuchadnezzar's name means: Nabo, protect the crown! Belshedezzar was the last king of Babylonia. He was the son of Nabonidus whose wife was Nictoris, the daughter of Nebuchadnezzar.

The first king of the first dynasty of Ur (around 2560) was Hammurabi. He made Ur Sumer's main city.

***First Dynasty of Ur c. 2563-2387 B.C.***

- 2563-2524: Mesannepadda
- 2523-2484: A'annepadda
- 2483-2448: Meskiagnunna
- 2447-2423: Elulu
- 2422-2387: Balulu

***Dynasty of Lagash c. 2494-2342 B.C.***

- 2494-2465: Ur-Nanshe
- 2464-2455: Akurgal
- 2454-2425: Ennatum
- 2424-2405: Enannatum I
- 2402-2375: Entemena
- 2374-2365: Enannatum II
- 2364-2359: Enentarzi
- 2358-2352: Lugal-anda
- 2351-2342: Uru-inim-gina

***Dynasty of Uruk c. 2340-2316 B.C.***

- 2340-2316: Lugal-zaggesi

***Dynasty of Akkad c. 2334-2154 B.C.***

- 2334-2279: Sargon
- 2278-2270: Rimush

## Power

When Assyria grew into an empire, it was divided into smaller parts, called provinces. Each of these were named after their main cities, like Nineveh, Samaria, Damascus and Arpad. They all had their own governor who had to make sure everyone paid their taxes; he had to call up soldiers to war, and supply workers when a temple was built. He was also responsible for the laws being enforced.

In this way it was easier to keep control of an empire like Assyria. Although Babylon was quite a small state in the Sumerian, it grew tremendously throughout the time of Hammurabi's rule. He was known as "the law maker", and soon Babylon became one of the main cities in Mesopotamia. It was later called Babylonia, which meant "the gateway of the gods." It also became one of history's greatest centers of learning.

## Warfare

As city-states began to grow, their spheres of influence overlapped, creating arguments between other city-states, especially over land and canals. These arguments were recorded in tablets several hundreds of years before any major war - the first recording of a war occurred around 3200BC but was not common until about 2500BC. At this point warfare was incorporated into the Mesopotamian political system, where a neutral city may act as an arbitrator for the two rival cities. This helped to form unions between cities, leading to regional states.

When empires were created, they went to war more with foreign countries. King Sargon, for example conquered all the cities of Sumer, some cities in Mari, and then went to war with northern Syria. Many Babylonian palace walls were decorated with the pictures of the successful fights and the enemy, whether desperately escaping, or hiding amongst reeds. A king in Sumer, Gilgamesh, was thought two-thirds god and only one third human. There were legendary stories and poems about him, which were passed on for many generations, because he had many adventures that were believed very important, and won many wars and battles.

## Laws

King Hammurabi, as mentioned above, was famous for his set of laws, The Code of Hammurabi (created ca. 1780 BC), which is one of the earliest sets of laws found and one of the best preserved examples of this type of document from ancient Mesopotamia. For more information, see Hammurabi and Code of Hammurabi.

## Architecture

The study of ancient Mesopotamian architecture is based on available archaeological evidence, pictorial representation of buildings and texts on building practices. Scholarly literature usually concentrates on temples, palaces, city walls and gates and other monumental buildings, but occasionally one finds works on residential architecture as well. Archaeological surface surveys also allowed for the study of urban form in early Mesopotamian cities.

Most notably known architectural remains from early Mesopotamia are the temple complexes at Uruk from the 4th millennium BC, temples and palaces from the Early Dynastic period sites in the Diyala River valley such as Khafajah and Tell Asmar, the Third Dynasty of Ur remains at Nippur (Sanctuary of Enlil) and Ur (Sanctuary of Nanna), Middle Bronze Age remains at Syrian-Turkish sites of Ebla, Mari, Alalakh, Aleppo and Kultepe, Late Bronze Age palaces at Bogazkoy (Hattusha), Ugarit, Ashur and Nuzi, Iron Age palaces and temples at Assyrian (Kalhu/Nimrud, Khorsabad, Nineveh), Babylonian (Babylon), Urartian (Tushpa/Van Kalesi, Cavustepe, Ayanis, Armavir, Erebuni, Bastam) and Neo-Hittite sites (Karkamis, Tell Halaf, Karatepe). Houses are mostly known from Old Babylonian remains at Nippur and Ur. Among the textual sources on building construction and associated rituals, Gudea's cylinders from the late 3rd millennium are notable, as well as the Assyrian and Babylonian royal inscriptions from the Iron Age.

## Houses

The materials used to build a Mesopotamian house were the same as those used today: mud brick, mud plaster and wooden doors, which were all naturally available round the city, although wood could not be naturally made very well during the particular time period described. Most houses had a square center room with other rooms attached to it, but a great variation in the size and materials used to build the houses suggest they were built by the inhabitants themselves. The smallest rooms may not have coincided with the poorest people; in fact it could be that the poorest people built houses out of perishable materials such as reeds on the outside of the city, but there is very little direct evidence for this.

## The Palace

The palaces of the early Mesopotamian elites were large scale complexes, and were often lavishly decorated. Earliest examples are known from the Diyala River valley sites such as Khafajah and Tell Asmar. These third millennium BC palaces functioned as a large scale socio-economic institutions, therefore, along with residential and private function, they housed craftsmen workshops, food storehouses, ceremonial courtyards, and often associated with shrines. For instance, the so-called "giparu" (or Gig-Par-Ku in Sumerian) at Ur where the Moon god Nanna's priestesses resided was a major complex with multiple courtyards, a number of sanctuaries, burial chambers for dead priestesses, a ceremonial banquet hall, etc. A similarly complex example of a Mesopotamian palace was excavated at Mari in Syria, dating from the Old Babylonian period.

Assyrian palaces of the Iron Age, especially at Kalhu/Nimrud, Dur Sharrukin/Khorsabad and Ninuwa/Nineveh, have become famous due to the pictorial and textual narrative programs on their walls, all carved on stone slabs known as orthostats. These pictorial programs either incorporated cultic scenes or the narrative accounts of the kings' military and civic accomplishments. Gates and important passageways were flanked with massive stone sculpture of apotropaic mythological figures.

The architectural arrangement of these Iron Age palaces were also organized around large and small courtyards. Usually the king's throneroom opened to a massive ceremonial courtyard where important state councils met, state ceremonies performed.

Massive amounts of ivory furniture pieces were found in many Assyrian palaces pointing out an intense trade relationship with North Syrian Neo-Hittite states at the time. There is also good evidence that bronze repousse bands decorated the wooden gates.

## Ziggurats

Ziggurats (Akkadian ziqquratu from the verb zaqaru) were massive stepped cult platforms found in certain Mesopotamian sanctuaries. The idea seems to have originated in early Mesopotamian temples which were built successively, one building over another on the same site over centuries, creating a massive mound that raised the new temples over the rest of the city.

A good example of such structure was the temple dedicated to Ea at Eridu (Tell Abu Shahrain) excavated by Fuad Safar and Seton Lloyd in 1940s, or the "White" Temple dedicated to Anu at Uruk in the Late Uruk period. Ur-Nammu's ziggurat, built at the height the Third Dynasty of Ur, at the site of Ur (Tell al Mugayyar) in the sanctuary of the Moon God Nanna, is also believed to be encasing earlier temples of the Early Dynastic Period.

Ur-Nammu's ziggurat is considered one of the earliest of all planned ziggurats. After that time Kassites and Elamites of the Late Bronze Age, and Assyrians and Babylonians of the Iron age continued to build artificially erected ziggurats. Examples of such structures were found in Dur Kurigalzu (Aqar Quf), Dur-Untash (Tschoga Zanbil), Kalhu (Nimrud), Dur-Sharrukin (Khorsabad) and Babylon among others.

It has been suggested that ziggurats were built to resemble mountains, but there is little textual or archaeological evidence to support that hypothesis.

Ur-Nammu's ziggurat at Ur was designed as a three-stage construction, today only two of these survive. This entire mudbrick core structure was originally given a facing of baked brick envelope set in bitumen, circa 2.5 m on the first lowest stage, and 1.15 m on the second. Each of these baked bricks were stamped with the name of the king. The sloping walls of the stages were buttressed. The access to the top was by means of a triple monumental staircase, which all converges at a portal that opened on a landing between the first and second stages. The height of the first stage was about 11 m while the second stage rose some 5.7 m. Usually a third stage is reconstructed by the excavator of the ziqqurat (Leonard Woolley), and crowned by a temple. At the Tschoga Zanbil ziggurat archaeologists have found massive reed ropes that ran across the core of the ziggurat structure and tied together the mudbrick mass.

### Economy

There was a large difference in money and wealth between rich and ordinary people. Ordinary people were highly dependent on their crops, because they had very little money. Rich people had many slaves and lots of money.

1 talent =

- 60 mina
- 3,600 shekel
- 30 kg of silver

1 mina =

- 60 shekel
- 500 grams of silver

1 shekel =

- 8.333 grams of silver
- 2 divisions
- 8 slices
- 12 grains
- 24 carats
- 24 chickpeas
- 180 barleycorns

Silver coins were not pure silver. About 87% of a coin was silver.

Much of the coinage was copper.

**More Recent History**

- The region was conquered by Cyrus the Great and came under the rule of the Persian Achaemenid Empire, apparently as two satrapies, Babylonia in the south and Athura (from Assyria) in the north. During this time, 500-330 BC, the Persian Empire became the pre-eminent power of the world.
- After the conquest of all Persia by the Hellenizing Macedonian king Alexander the Great, the satrapies were part of the major diadochy, the Seleucid Empire, until just before its elimination by Greater Armenia in 42 BC.
- Most of Mesopotamia then became part of the Parthian Empire of Persia, which lasted until 224 AD. Ctesiphon was made the capital of the Parthian Empire. However, a part in the northwest became Roman. Under the Tetrarchy this part was divided into two provinces: Osrhoene (around Edessa, roughly the modern-day border between Turkey and Syria) and Mesopotamia (a bit more northeast).
- During the time of the Persian Sassanid Empire, the much larger share of Mesopotamia was called Del-e Iranshahr meaning "Iran's Heart" and the metropol Ctesiphon (facing ancient Seleukia across the Tigris), the capital of Persia, was situated in Mesopotamia.
- In the early 7th century AD, the Muslim Arab tribes under Khalid ibn al-Walid conquered all of the Persian Sassanid Empire during the Islamic conquest of Persia, and the Arab Empire was soon established under the Rashidun and Umayyad Caliphates. Consequently, Mesopotamia was reunited under the Arabs, but governed as two provinces: northern, with Mosul (also known as Nineveh) as its capital, and southern, with Baghdad as its capital. Later under the Abbasid caliphs, Baghdad became the capital of the Arab Empire until the sack of Baghdad in 1258.
- From 1508-1534 AD, the Persian Safavids took control of Mesopotamia.

- In 1535 AD, Ottoman Turks took over Baghdad. During the reign of the Ottoman Empire, Mesopotamia was ruled as three separate vilayats, or territories: Mosul, Baghdad, and Basra, which included the territory that is now present-day Kuwait.
- At the end of World War I Mesopotamia was briefly occupied by the British, who, under the authority of the British Mandate of Mesopotamia, set up the government of what is now present day Syria and Iraq under one Hashemite ruler.
- In 1920 the nation-state of Iraq was formed by the United Kingdom following the dissolution of the Ottoman Empire, with its present-day borders and including the territory that is now known as Kuwait. Kuwait, a British protectorate, which had originally been a part of the Basra province under Ottoman rule, was granted independence from Britain in 1961.
- In the early 1990s coalition forces launched an attack on Iraq known as Operation Desert Storm, in response to President Saddam Hussein's invasion of Kuwait.
- A German archaeologist says he has found relics of "humanity's first war" in the north east of Syria in the form of balls of stone used as ammunition in the 4th millenium BC, the Die Zeit newspaper says in its edition due for publication on Thursday, January 4, 2007. "We have there the oldest example of an offensive war," said Clemens Reichel, who is leading an archaeological dig in the ancient city of Hamoukar, on the border with Iraq, for the University of Chicago. Reichel said that almost 6,000 years ago the city, whose fortifications were three metres (ten feet) thick, was besieged and reduced to ashes probably by attackers from southern Mesopotamia.

## SUMERIAN ARCHITECTURE

The Tigris-Euphrates plain lacked minerals and trees. Sumerian structures comprised plano-convex mudbrick, not fixed with mortar nor cement. As plano-convex bricks (being rounded) are somewhat unstable in behaviour, Sumerian bricklayers would lay a row of bricks perpendicular to the rest every few rows. They would fill the gaps with bitumen, straw, marsh reeds, and weeds.

Mud-brick buildings eventually deteriorate, so they were periodically destroyed, levelled, and rebuilt on the same spot. This constant rebuilding gradually raised the level of cities, so that they came to be elevated above the surrounding plain. The resulting hills are known as tells, and are found throughout the ancient Near East. Sumerian cylinder seals also depict houses built from reeds, not unlike those built by the Marsh Arabs of Southern Iraq until recent years.

Sumerian temples and palaces made use of more advanced materials and techniques, such as buttresses, recesses, half columns, and clay nails.

Scribes were also important to Sumerian architecture, to make records of construction carried out for government, nobility, or royalty.

### Ziggurats

The most famous Sumerian buildings are the ziggurats - large terraced platforms with temples on top. Such ziggurats may have been the inspiration for the Biblical Tower of Babel (see Etemenanki).

Ziggurats typical of the Ubaid period were built very high on a platform of mud brick. On these large platforms were built gradually smaller and smaller concentric platforms, although sometimes there were ground level temples more typical of the protoliterate period. These were similar to some modern buildings in the shape of ziggurats. Many temples had inscriptions engraved into them, such as the one at Uqair.

## ART AND ARCHITECTURE OF ASSYRIA

In Babylonia, an abundance of clay, and lack of stone, led to greater use of mudbrick; Babylonian temples are massive structures of crude brick, supported by buttresses, the rain being carried off by drains. One such drain at Ur was made of lead. The use of brick led to the early development of the pilaster and column, and of frescoes and enamelled tiles. The walls were brilliantly coloured, and sometimes plated with zinc or gold, as well as with tiles. Painted terra-cotta cones for torches were also embedded in the plaster.

Assyria, imitating Babylonian architecture, also built its palaces and temples of brick, even when stone was the natural building material of the country - faithfully preserving the brick platform, necessary in the marshy soil of Babylonia, but little needed in the north.

As time went on, however, later Assyrian architects began to shake themselves free of Babylonian influence, and to use stone as well as brick. The walls of Assyrian palaces were lined with sculptured and coloured slabs of stone, instead of being painted as in Chaldea. Three stages may be traced in the art of these bas-reliefs: it is vigorous but simple under Ashurnasirpal II, careful and realistic under Sargon II, and refined but wanting in boldness under Ashurbanipal.

In Babylonia, in place of the bas relief, there is greater use of three-dimensional figures in the round - the earliest examples being the statues from Telloh, that are realistic if somewhat clumsy. The paucity of stone in Babylonia made every pebble precious, and led to a high perfection in the art of gem-cutting.

Two seal-cylinders from the age of Sargon of Akkad are among the best examples of their kind. One of the first remarkable specimens of early metallurgy to be discovered by archaeologists is the silver vase of Entemena. At a later epoch, great excellence was attained in the manufacture of such jewellery as earrings and bracelets of gold. Copper, too, was worked with skill; indeed, it is possible that Babylonia was the original home of copper-working. The people were famous at an early date for their embroideries and rugs.

The forms of Assyrian pottery are graceful; the porcelain, like the glass discovered in the palaces of Nineveh, was derived from Egyptian models. Transparent glass seems to have been first introduced in the reign of Sargon. Stone, clay and glass were used to make vases, and vases of hard stone have been dug up at Telloh similar to those of the early dynastic period of Egypt.

## SOCIAL LIFE IN BABYLONIA AND ASSYRIA

The priesthood of Babylonia was divided into a great number of classes, including a medicinal class. It had a counterpart in the military aristocracy of Assyria. The army was raised by conscription; the concept of a standing army seems to have been first organized in Assyria. Successive improvements were introduced into it by the kings of the second Assyrian empire; chariots were replaced by cavalry; Tiglath-Pileser III gave the riders saddles and high boots, and Sennacherib created a corps of slingers. Tents, baggage-carts and battering-rams were carried on the march, and the tartan or commander-in-chief ranked next to the king.

In both countries, there was a large body of slaves; above them came the agriculturists and commercial classes, who were comparatively few in Assyria. The scribes, on the other hand, were a more important class in Assyria than in Babylonia. Both countries had their artisans, money-lenders, poets and musicians. The houses of the people contained little furniture; chairs, tables and couches were used, and Ashurbanipal is represented as reclining on his couch at a meal while his wife sits on a chair beside him.

After death, the body was usually partially cremated, along with the objects that had been buried with it. The cemetery adjoined the city of the living, and was laid out in streets through which ran rivulets of "pure" water. Many of the tombs, built of crude brick, were provided with gardens, and there were shelves or altars with offerings to the dead. As the older tombs decayed, a fresh city of tombs arose on their ruins. It is remarkable that thus far, no cemetery older than the Seleucid or Parthian period has been found in Assyria.

### Babylonian Law

The material for the study of Babylonian law is singularly extensive. The so-called "contracts" exist in the thousands, including a great variety of deeds, conveyances, bonds, receipts, accounts, and most important of all, the actual legal decisions given by the judges in the law courts. Historical inscriptions, royal charters and rescripts, dispatches, private letters and the general literature afford welcome supplementary information.

Even grammatical and lexicographical works contain many extracts or short sentences bearing on law and custom. The so-called "Sumerian Family Laws" are thus preserved.

The discovery of the now-celebrated Code of Hammurabi (hereinafter simply termed "the Code") has made a more systematic study possible than could have resulted from the classification and interpretation of the other material. Some fragments of a later code exist and have been published; but there still remain many points whereof we have no evidence.

This material dates from the earliest times up to the commencement of the common era. Evidence on a particular point may be very full at one period, and almost entirely lacking for another. The Code forms the backbone of the skeleton sketch that is here reconstructed. The fragments of it that have been recovered from Assur-bani-pal's library at Nineveh and later Babylonian copies show that it was studied, divided into chapters, entitled Ninu ilu sirum from its opening words, and recopied for fifteen hundred years or more. The greater part of it remained in force, even through the Persian, Greek and Parthian conquests, that affected private life in Babylonia very little; and it survived to influence Syro-Roman and later Islamic law in Mesopotamia. The laws and customs that preceded the Code, we shall call "early"; that of the Neo-Babylonian empire (as well as the Persian, Greek, etc.) "late." The law of Assyria was derived from Babylonia, but conserved early features long after they had disappeared elsewhere.

## Tribal Influences

When the Semitic tribes settled in the cities of Mesopotamia, their tribal customs passed over into city law. The early history of the country is the story of a struggle for supremacy between the cities. A metropolis demanded tribute and military support from its subject cities, but left their local cults and customs unaffected. The city rights and usages were respected by kings and conquerors alike.

As late as the accession of Assur-bani-pal and Shamash-shum-ukin, we find the Babylonians appealing to their city laws that groups of aliens to the number of twenty at a time were free to enter the city; that foreign women, once married to Babylonian husbands, could not be enslaved; and that not even a dog that entered the city could be put to death untried.

The population of Babylonia was of many races from early times, and intercommunication between the cities was incessant. Every city had a large number of resident aliens. This freedom of intercourse must have tended to assimilate custom. It was, however, reserved for the genius of Hammurabi to make Babylon his metropolis and weld together his vast empire by a uniform system of law.

Almost all trace of tribal custom had already disappeared from the law of the Code. It is state-law; self-help, blood-feud, marriage by capture, are all absent; though the Code of family solidarity, district responsibility, ordeal, the lex murabi, talionis, are primitive features that remain. The king is a benevolent autocrat, easily accessible to all his subjects, both able and willing to protect the weak against the highest-placed oppressor. The royal power, however, can only pardon when private resentment is appeased. Judges are strictly supervised, and appeal is allowed. The whole land is covered with feudal holdings, masters of the levy, police, etc. There is a regular postal system. The pax Babylonica is so assured that private individuals do not hesitate to ride in their carriage from Babylon to the coast of the Mediterranean. The position of women is free and dignified.

The Code did not merely embody contemporary custom or conserve ancient law. It is true that centuries of law-abiding and litigious habitude had accumulated, in the temple archives of each city, vast stores of precedent in ancient deeds and records of judicial decisions, and that intercourse had assimilated city custom. The universal habit of writing, and perpetual recourse to written contract, even more modified primitive custom and ancient precedent.

Provided the parties could agree, the Code left them free to contract, as a rule. Their deed of agreement was drawn up in the temple by a notary public, and confirmed with an oath "by god and the king." It was publicly sealed, and witnessed by professional witnesses, as well as by collaterally interested parties. The manner whereby it was thus executed may have been sufficient guarantee that its stipulations were not impious or illegal. Custom or public opinion doubtless secured that the parties would not agree to "wrong." In case of dispute, the judges dealt first with the contract. They might not sustain it, but if the parties did not dispute it, they were free to observe it. The judges' decision might, however, be appealed against. Many contracts contain the proviso that in case of future dispute, the parties would abide by "the decision of the king." The Code made known, in a vast number of cases, what that decision would be, and many cases of appeal to the king were returned to the judges with orders to decide in accordance with it. The Code itself was carefully and logically arranged, and the order of its sections was conditioned by their subject-matter. Nevertheless, the order is not that of modern scientific treatises, so a somewhat different order from both is most convenient for our purpose.

**Three Classes**

The Code contemplates the whole population as falling into three classes: the amelu, the muskinu and the ardu.

The amelu was a patrician, the man of family, whose birth, marriage and death were registered; of ancestral estates and full civil rights. He had aristocratic privileges and responsibilities, and the right to exact retaliation for corporal injuries, but was liable to a heavier punishment for crimes and misdemeanours, higher fees and fines. To this class belonged the king and court, the higher officials, the professions and craftsmen. The term became a mere courtesy title over time, but originally carried with it a certain status. Already in the Code, when status is not concerned, it is used to denote "anyone." There was no property qualification, nor does the term appear to be racial.

It is most difficult to characterize the muskinu exactly. The term came in time to mean "a beggar", and with that meaning has passed through Aramaic and Hebrew into many modern languages; but though the Code does not regard him as necessarily poor, he may have been landless. He was free, but had to accept monetary compensation for corporal injuries, paid smaller fees and fines, even paid less offerings to the gods. He inhabited a separate quarter of the city. There is no reason to regard him as specially connected with the court, as a royal pensioner, nor as forming the bulk of the population. The rarity of any references to him in contemporary documents makes further specification conjectural.

The ardu was a slave, his master's chattel, and formed a very numerous class. He could acquire property and even hold other slaves. His master clothed and fed him, and paid his doctor's fees, but took all compensation paid for injury done to him. His master usually found him a slave-girl as wife (the children were then born slaves), often set him up in a house (with farm or business) and simply took an annual rent of him. Otherwise, he might marry a free woman (the children were then free), who might bring him a dower that his master could not touch, and at his death, one-half of his property passed to his master as his heir. He could acquire his freedom by purchase from his master, or might be freed and dedicated to a temple, or even adopted, when he became an amelu and not a muskinu. Slaves were recruited by purchase abroad, from captives taken in war, or by freemen degraded for debt or crime. A slave often ran away; if caught, the captor was bound to restore him to his master, and the Code fixes a reward of two shekels that the owner must pay the captor. It was about one-tenth of the average value. To detain or harbour, etc., a slave was punishable by death. So was an attempt to get him to leave the city. A slave bore an identification mark, removable only by a surgical operation, and that later consisted of his owner's name tattooed or branded on the arm. On the great estates in Assyria and its subject provinces, there were many serfs, mostly of subject race, settled captives, or quondam slaves; tied to the soil they cultivated, and sold with the estate, yet capable of possessing land and property of their own. There is little trace of serfs in Babylonia, unless the muskinu be really a serf.

### Citizens Tenants of Gods

The god of a city was originally considered the owner of its land, that encircled it with an inner ring of irrigable arable land and an outer fringe of pasture; and the citizens were his tenants. The god and his vice regent, the king, had long ceased to disturb tenancy, and were content with fixed dues in naturalia, stock, money or service.

One of the earliest monuments records the purchase by a king of a large estate for his son, paying a fair market price and adding a handsome honorarium to the many owners, in costly garments, plate, and precious articles of furniture. The Code recognizes complete private ownership of land, but apparently extends the right to hold land to votaries, merchants (and resident aliens?). But all land was sold subject to its fixed charges. The king, however, could free land from these charges by charter, which was a frequent way of rewarding those who deserved well of the state.

It is from these charters that we learn nearly all we know of the obligations lying upon land. The state demanded men for the army and the corvée, as well as dues in kind. A defined area was bound to find a bowman, together with his linked pikeman (who bore the shield for both), and to furnish them with supplies for the campaign. This area was termed a "bow" as early as the 8th century BC, but the practice was much earlier. Later, a horseman was also due from certain areas. A man was only bound to serve so many (six?) times, but the land still had to find a man annually. This service was usually discharged by slaves and serfs, but the amelu (and perhaps the muskinu) also went to war. The "bows" were grouped in tens and hundreds. The corvée was less regular. The letters of Hammurabi often deal with claims to exemption. Religious officials and shepherds in charge of flocks were exempt.

Special liabilities lay upon riparian owners to repair canals, bridges, quays, etc. The state claimed certain proportions of all crops, stock, etc. The king's messengers could commandeer any subject's property, giving a receipt. Further, every city had its own octroi duties, customs, ferry dues, highway and water rates. The king had long ceased to be owner of the land, if he ever was. He had his own royal estates, his private property, and dues from all his subjects. The higher officials had endowments and official residences.

The Code regulates the feudal position of certain classes. They held an estate from the king, consisting of house, garden, field, stock, and a salary, on condition of personal service on the king's errand. They could not delegate the service, on penalty of death. When ordered abroad, they could nominate a capable son to hold the benefice and carry on the duty. If there were no capable son, the state put in a locum tenens, but granted one-third to the wife to maintain herself and children. The fief was otherwise inalienable; it could not be sold, pledged, exchanged, sublet, devised or diminished. Other land was leased from the state. Ancestral estate was strictly tied to the family. If a holder would sell, the family kept the right of redemption, and there seems to have been no time-limit to its exercise.

### Temple

The temple occupied a most important position. It received from its estates, from tithes and other fixed dues, as well as from the sacrifices (a customary share) and other offerings of the faithful, vast amounts of all sorts of naturalia; besides money and permanent gifts. The larger temples had many officials and servants.

Originally, perhaps, each town clustered round one temple, and each head of a family had a right to minister there and share its receipts. As the city grew, the right to so many days a year at one or other shrine (or its "gate") descended within certain families, and became a kind of property that could be pledged, rented or shared within the family, but not alienated. Despite all these demands, the temples became great granaries and store-houses, as they were also the city archives. The temple had its responsibilities. If a citizen were captured by the enemy and could not ransom himself, the temple of his city must do so. To the temple came the poor farmer to borrow seed, grain, or supplies for harvesters, etc. - advances that he repaid without interest.

The king's power over the temple was not proprietary, but administrative. He might borrow from it, but repaid like other borrowers. The tithe seems to have been considered the rent due to the god for his land. It is not clear that all lands paid tithe; perhaps only such as once had a special connection with the temple. The Code deals with a class of persons devoted to the service of a god, as vestals or hierodules. The vestals were vowed to chastity, lived together in a great nunnery, were forbidden to enter a tavern, and, together with other votaries, had many privileges.

## Property Law

The Code recognizes many ways of disposing of property: sale, lease, barter, gift, dedication, deposit, loan, or pledge, all of which were matters of contract. "Sale" was the delivery of the purchase (in the case of real estate, symbolized by a staff, a key, or deed of conveyance) in return for the purchase money, receipts being given for both. Credit, if given, was treated as a debt, and secured as a loan by the seller to be repaid by the buyer, for which he gave a bond.

The Code admits no claim unsubstantiated by documents or the oath of witnesses. A buyer had to convince himself of the seller's title. If he bought (or received on deposit) from a minor or a slave without power-of-attorney, he would be executed as a thief. If the goods were stolen and the rightful owner reclaimed them, he had to prove his purchase by producing the seller and the deed of sale, or witnesses to it; otherwise he would be adjudged a thief and die. If he proved his purchase, he had to give up the property but had his remedy against the seller or, if he had died, could reclaim fivefold from his estate.

A man who bought a slave abroad, might find that he had previously been stolen or captured from Babylonia, and he then had to restore him to his former owner without recompense. If he bought property belonging to a feudal holding, or to a ward in chancery, he had to return it and forfeit what he gave for it as well. He could repudiate the purchase of a slave attacked by the bennu sickness within the month (later, a hundred days), and held a newly-purchased female slave three days "on approval". A defect of title, or an undisclosed liability, would invalidate a sale at any time.

## Leasing

Landowners frequently cultivated their land themselves, but might employ a husbandman, or rent it. The husbandman was bound to carry out the proper cultivation, raise an average crop, and leave the field in good tilth. In case the crop failed, the Code fixed a statutory return. Land might be leased at a fixed rent, when the Code enacted that accidental loss fell on the tenant. If let on share-profit, the landlord and tenant shared the loss proportionately to their stipulated share of profit. If the tenant paid his rent and left the land in good tilth, the landlord could not interfere nor forbid subletting.

Waste land was let to reclaim, the tenant being rent-free for three years and paying a stipulated rent in the fourth year. If the tenant neglected to reclaim the land, the Code enacted that he must hand it over in good tilth and fixed a statutory rent. Gardens or plantations were leased in the same ways and under the same conditions; but for date groves, four years' free tenure was allowed.

The metayer system was in vogue, especially on temple lands. Thc landlord found land, labour, oxen for ploughing and working the watering machines, carting, threshing or other implements, grain seed, rations for the workmen and fodder for the cattle.

The tenant, or steward, usually had other land of his own. If he stole the seed, rations or fodder, the Code enacted that his fingers be cut off. If he appropriated or sold the implements, impoverished or sublet the cattle, he was heavily fined, and in default of payment might be condemned to be torn to pieces by the cattle on the field. Rent was as contracted.

Irrigation was indispensable. If the irrigator neglected to repair his dyke, or left his runnel open and caused a flood, he had to make good the damage done to his neighbours' crops, or be sold with his family to pay the cost. The theft of a watering machine, water-bucket or other agricultural implement was heavily fined.

Houses were usually leased for the year, but also for longer terms, rent being paid in advance, half-yearly. The contract generally specified that the house be in good repair, and the tenant was bound to keep it so.

The woodwork, including doors and door frames, was removable, and the tenant might bring and take away his own. The Code enacted that if the landlord would re-enter before the term was up, he must remit a fair proportion of the rent. Land was leased for houses or other buildings to be built upon it, the tenant being rent-free for eight or ten years; after which the building came into the landlord's possession.

## Hired Labour

Despite the multitude of slaves, hired labour was often needed, especially at harvest. This was a matter of contract, and the employer, who usually paid in advance, might demand a collateral against fulfillment of the work. Cattle were hired for ploughing, working the watering-machines, carting, threshing, etc. The Code fixed a statutory wage for sowers, ox-drivers, field-labourers, and hire for oxen, asses, etc.

There were many herds and flocks. The flocks were committed to a shepherd, who gave receipt for them and took them out to pasture. The Code fixed his wage. He was responsible for all care, must restore ox for ox, sheep for sheep, must breed them satisfactorily. Any dishonest use of the flock had to be repaid tenfold, but loss due to disease or wild beasts fell upon the owner. The shepherd made good all loss due to his own neglect. If he let the flock feed on a field of crops, he had to pay damages fourfold; if he turned them into standing crops when they ought to have been folded, he paid twelvefold.

## Debt

In commercial matters, payment in kind was still common, though the contracts usually stipulate for cash, naming the standard expected--that of Babylon, Larsa, Assyria, Carchemish, etc. The Code enacted, however, that a debtor must be allowed to pay in produce according to a statutory scale. If a debtor had neither money nor crops, the creditor must not refuse goods.

Debt was secured on the person of the debtor. Distraint on a debtor's grain was forbidden by the Code; not only must the creditor return it, but his illegal action forfeited his claim altogether. An unwarranted seizure for debt was fined, as was the distraint of a working ox.

The debtor being seized for debt could nominate as mancipium, or hostage to work off the debt, his wife, a child, or slave. The creditor could only hold a wife or child three years as mancipium. If the mancipium died a natural death while in the creditor's possession, no claim could lie against the latter; but if he was the cause of death by cruelty, he had to give son for son, or pay for a slave. He could sell a slave-hostage, apart from a slave-girl who had borne her master children; she had to be redeemed by her owner.

The debtor could also pledge his property, and in contracts often pledged a field, house or crop. The Code enacted, however, that the debtor should always take the crop himself and pay the creditor from it. If the crop failed, payment was deferred, and no interest could be charged for that year. If the debtor did not cultivate the field himself, he had to pay for the cultivation, but if the cultivation was already finished, he must harvest it himself and pay his debt from the crop. If the cultivator did not get a crop, this would not cancel his contract.

Pledges were often made where the intrinsic value of the article was equivalent to the amount of the debt; but antichretic pledge was more common, where the profit of the pledge was a set-off against the interest of the debt. The whole property of the debtor might be pledged as collateral for the payment of the debt, without any of it coming into the enjoyment of the creditor. Personal guarantees were often given in Babylon that the debtor would repay, or the guarantor become liable himself.

**Trade**

Trade was very extensive. A common procedure was for a merchant to entrust his goods or money to a travelling agent, who sought a market for his goods. The caravans travelled far beyond the limits of the empire.

The Code insisted that the agent should inventory and give a receipt for all that he received. No claim could be made for anything not so entered. Even if the agent made no profit, he was bound to return double what he had received; if he made poor profit, he had to make up the deficiency; but he was not responsible for loss by robbery or extortion on his travels. On his return, the lending merchant must give him a receipt for what was handed over to him. Any false entry or claim on the agent's part was penalised threefold; on the lending merchant's part, sixfold. In normal cases, profits were divided according to contract, usually equally.

A considerable amount of forwarding (advancing wares to the agent up front) was done by the caravans. The carrier gave a receipt for the consignment, took all responsibility, and exacted a receipt upon delivery. If he defaulted, he paid fivefold. He was usually paid in advance. Deposit, especially warehousing of grain, was charged for at one-sixtieth. The warehouse man took all risks, paid double for all shortage, but no claim could be made unless he had given a properly witnessed receipt.

Water traffic on the Euphrates and canal system was early on, quite considerable. Ships, whose tonnage was estimated by the amount of grain they could carry, were continually hired for the transport of all kinds of goods. The Code fixes the price for shipbuilding, and insists on the builder's giving a year's guarantee of seaworthiness. It also fixes the rate of hire for ship and crew. The captain was responsible for the freight and the ship; he had to replace all loss. Even if he refloated the ship, he had to pay a fine of half its value for sinking it. In the case of collision, the boat under way was responsible for damages to the boat at anchor.

The Code also regulated the liquor traffic - fixing a fair price for beer, and forbidding the connivance of the tavern-keeper (a female) at disorderly conduct or treasonable assembly, under pain of death. She was required to take the offenders to the palace - implying an efficient and accessible police system.

Payment through a banker or by written draft against deposit was frequent. Bonds to pay were treated as negotiable. Interest was rarely charged on advances by the temple or wealthy landowners for pressing needs, but this may have been part of the metayer system. The borrowers may have been tenants. Interest was charged at very high rates for overdue loans of this kind. Merchants (and even temples in some cases) made ordinary business loans, charging from 20% to 30%.

## FAMILY LAW

### Marriage

Marriage retained the form of purchase, but was essentially a contract to be man and wife together. The marriage of young people was usually arranged between the relatives - the groom's father providing the bride-price, which with other presents, the suitor ceremonially presented to the bride's father. This bride-price was usually handed over by her father to the bride upon her marriage, and so returned into the bridegroom's possession, along with her dowry, which was her portion as a daughter.

The bride-price varied greatly, according to the status of the parties, but surpassed the price of a slave. The Code enacted that if the father does not, after accepting a man's presents, give him his daughter, he must return the presents doubled. This was done even if his decision was brought about by libel on the part of the suitor's friend, and the Code enacted that the faithless friend should not marry the girl. If a suitor changed his mind, he forfeited the presents.

The dowry might include real estate, but generally consisted of personal effects and household furniture. It remained the wife's for life, descending to her children, if any; otherwise returning to her family, when the husband could deduct the bride-price if it had not been given to her, or return it, if it had.

The marriage ceremony included joining of hands and the utterance of some formula of acceptance on the part of the bridegroom, as "I am the son of nobles, silver and gold shall fill thy lap, thou shalt be my wife, I will be thy husband. Like the fruit of a garden I will give thee offspring." It must be performed by a freeman.

The marriage contract - without which, the Code ruled that the woman was no wife - usually stated the consequences to which each party was liable for repudiating the other. These by no means necessarily agree with the Code. Many conditions might be inserted: as that the wife should act as maidservant to her mother-in-law, or to a first wife.

The married couple formed a unit as to external responsibility, especially for debt. The man was responsible for debts contracted by his wife, even before her marriage, as well as for his own; but he could use her as a mancipium. Hence the Code allowed a proviso to be inscribed in the marriage contract, that the wife should not be seized for her husband's pre-nuptial debts; but enacted that then he was not responsible for her pre-nuptial debts, and, in any case, that both together were responsible for all debts contracted after marriage. A man might make his wife a settlement by deed of gift, which gave her a life interest in part of his property, and he might reserve to her the right to bequeath it to a favourite child; but she could in no case leave it to her family. Although married, she always remained a member of her father's house - she is rarely named wife of A; usually daughter of B, or mother of C.

## Divorce

Divorce was optional with the man, but he had to restore the dowry, and if the wife had borne him children, she had the custody of them. He had then to assign her the income of field, or garden, as well as goods, to maintain herself and children until they grew up. She then shared equally with them in the allowance (and apparently in his estate at his death) and was free to marry again. If she had no children, he returned her the dowry, and paid her a sum equivalent to the bride-price - or a mina of silver, if there had been none. The latter is the forfeit usually named in the contract for his repudiation of her.

If she had been a bad wife, the Code allowed him to send her away, while he kept the children and her dowry; or he could degrade her to the position of a slave in his own house, where she would have food and clothing. She might bring an action against him for cruelty and neglect and, if she proved her case, obtain a judicial separation, taking with her, her dowry. No other punishment fell on the man. If she did not prove her case, but was proved to be a bad wife, she was drowned.

If she were left without maintenance during her husband's involuntary absence, she could cohabit with another man, but must return to her husband if he came back, the children of the second union remaining with their own father. If she had maintenance, a breach of the marriage tie was adultery. Wilful desertion by, or exile of, the husband dissolved the marriage, and if he returned, he had no claim on her property; possibly not on his own.

## Widowhood

A widow took her husband's place in the family - living on in his house and bringing up the children. She could only remarry with judicial consent, when the judge was bound to inventory the deceased's estate and hand it over to her and her new husband in trust for the children. They could not alienate a single utensil.

If she did not remarry, she lived on in her husband's house and took a child's share on the division of his estate, when the children had grown up. She still retained her dowry and any settlement deeded to her by her husband. This property came to her children. If she had remarried, all her children shared equally in her dowry, but the first husband's gift fell to his children, or to her selection among them, if so empowered.

### Childbearing

Monogamy was the rule, and a childless wife might give her husband a maid (who was no wife) to bear him children, who were then reckoned hers. She remained mistress of her maid, and might degrade her to slavery again for insolence, but could not sell her if she had borne her husband children. If the wife did this, the Code did not allow the husband to take a concubine; but if she would not, he could do so. The concubine was a co-wife, though not of the same rank; the first wife had no power over her. A concubine was a free woman, was often dowered for marriage, and her children were legitimate. She could only be divorced on the same conditions as a wife. If a wife became a chronic invalid, the husband was bound to maintain her in the home they had made together, unless she preferred to take her dowry and return to her father's house; but he was free to remarry. In all these cases, the children were legitimate and lawful heirs.

There was, of course, no hindrance to a man having children by a slave girl. These children were free in any case, and their mother then could not be sold, though she might be pledged, and she became free upon her master's death. Her children could be legitimized by their father's acknowledgment before witnesses, and were often adopted. They then ranked equally in sharing their father's estate; but if not adopted, the wife's children divided and took first choice.

Vestal virgins were not supposed to have children, yet they could marry and often did. The Code contemplated that such a wife would give a husband a maid, as above.

Free women might marry slaves and still be dowered for the marriage. The children were free, and at the slave's death, the wife took her dowry and half of what she and her husband had acquired in wedlock for self and children; the master taking the other half, as his slave's heir.

A father had control over his children until their marriage. He had a right to their labour, in return for their keep. He might hire them out and receive their wages, pledge them for debt, even sell them outright. Mothers had the same rights in the absence of the father; elder brothers, when both parents were dead. A father had no claim on his married children for support, but they retained the right to inherit on his death.

The daughter was not only in her father's power to be given in marriage, but he might dedicate her to the service of some god as a vestal or a hierodule; or give her as a concubine. She had no choice in these matters, often decided in her childhood. A grown-up daughter might wish to become a votary, perhaps in preference to an uncongenial marriage, and it seems that her father could not refuse her wish.

In all these cases, the father might dower her. If he did not, on his death the brothers were bound to do so, giving her a full child's share if a wife, a concubine or a vestal, but one-third of a child's share if she were a hierodule or a Marduk priestess. The latter had the privilege of exemption from state dues, and absolute disposal of her property. All other daughters had only a life interest in their dowry, which reverted to their family, if childless, or went to their children if they had any. A father might, however, execute a deed granting a daughter power to leave her property to a favourite brother or sister. A daughter's estate was usually managed for her by her brothers, but if they dissatisfied her, she could appoint a steward. If she married, her husband managed it. The son also appears to have received his share on marriage, but did not always then leave his father's house; he might bring his wife there. This was usual in child marriages.

## Adoption

Adoption was very common, especially where the father (or mother) was childless, or had seen all his children grow up and marry away. The child was then adopted to care for the parents' old age. This was done by contract, that usually specified what the parent had to leave and what maintenance was expected. The real children, if any, were usually consenting parties to an arrangement that cut off their expectations. They even, in some cases, found the estate for the adopted child who was to relieve them of care. If the adopted child failed to carry out the filial duty, the contract was annulled in the law courts. Slaves were often adopted, and if they proved unfilial, were reduced to slavery again.

A craftsman often adopted a son to learn the craft. He profited by the son's labour. If he failed to teach his son the craft, that son could prosecute him and get the contract annulled. This was a form of apprenticeship, and it is not clear that the apprentice had any filial relation.

A man who adopted a son, and afterwards married and had a family of his own, could dissolve the contract, and must give the adopted child one-third of a child's share in goods, but no real estate. That could only descend in his former family. Vestals frequently adopted daughters, usually other vestals, to care for their old age.

Adoption had to be with consent of the real parents, who usually executed a deed making over the child, who thus ceased to have any claim upon them. But vestals, hierodules, certain palace officials and slaves had no rights over their children and could raise no obstacle. Orphans and illegitimate children had no parents to object. If the adopted child discovered his true parents and wanted to return to them, his eye or tongue was torn out. An adopted child was a full heir; the contract might even assign him the position of eldest son. Usually he was residuary legatee.

**Heirs**

All legitimate children shared equally in the father's estate at his death, reservation being made of a bride-price for an unmarried son, dower for a daughter, or property deeded to favourite children by the father. There was no birthright attaching to the position of eldest son, but he usually acted as executor, and after considering what each had already received, equalized the shares. He even made grants in excess to the others from his own share. When there were two widows with legitimate issue, both families shared equally in the father's estate, until later times, when the first family took two-thirds. Daughters, in the absence of sons, had sons' rights. Children also shared their own mother's property, but had no share in that of a stepmother.

A father could disinherit a son in early times without restriction, but the Code insisted upon judicial consent, and that only for repeated unfilial conduct. In early times, the son who denied his father had his front hair shorn, a slave-mark put on him, and could be sold as a slave; while if he denied his mother he had his front hair shorn, was driven round the city as an example and expelled from his home, but not degraded to slavery.

## Adultery

Adultery was punished with the death of both parties by drowning; but if the husband was willing to pardon his wife, the king might intervene to pardon the paramour. For incest with his own mother, both were burned to death; with a stepmother, the man was disinherited; with a daughter, the man was exiled; with a daughter-in-law, he was drowned; with a son's fiancee, he was fined. A wife who for her lover's sake procured her husband's death was gibbeted. A betrothed girl, seduced by her prospective father-in-law, took her dowry and returned to her family, and was free to marry as she chose.

## Punishment

In the criminal code, the ruling principle was the lex talionis. Eye for eye, tooth for tooth, limb for limb was the penalty for assault upon an amelu. A sort of symbolic retaliation was the punishment of the offending member, seen in the cutting off the hand that struck a father or stole a trust; in cutting off the breast of a wet nurse who substituted a changeling for the child entrusted to her; in the loss of the tongue that denied father or mother (in the Elamite contracts, the same penalty was inflicted for perjury); in the loss of the eye that pried into forbidden secrets. The loss of the surgeon's hand that caused loss of life or limb; or the brander's hand that obliterated a slave's identification mark, are very similar. The slave who struck a freeman or denied his master, lost an ear, the organ of hearing and symbol of obedience. To bring another into danger of death by false accusation was punished by death. To cause loss of liberty or property by false witness was punished by the same penalty the perjurer sought to bring upon another.

The death penalty was freely awarded for theft, and other crimes regarded as coming under that section: for theft involving entrance of palace or temple treasury, for illegal purchase from minor or slave, for selling stolen goods or receiving the same, for common theft in the open (in default of multiple restoration) or receiving the same, for false claim to goods, for kidnapping, for assisting or harbouring fugitive slaves, for detaining or appropriating same, for brigandage, for fraudulent sale of drink, for disorderly conduct of tavern, for delegation of personal service, for misappropriating the levy, for oppression of feudal holders, for causing death of a householder by bad building. The manner of death is not specified in these cases.

This death penalty was also fixed for such conduct as placed another in danger of death. A specified form of death penalty occurs in the following cases: gibbeting (on the spot where crime was committed) for burglary, later also for encroaching on the king's highway, for getting a slave-brand obliterated, for procuring a husband's death; burning for incest with own mother, for a vestal entering or opening a tavern, for theft at fire (on the spot); drowning for adultery, rape of a betrothed maiden, bigamy, bad conduct as wife, seduction of a daughter-in-law.

A curious extension of the talio is the death of a creditor's son for his father's having caused the death of a debtor's son as mancipium; of a builder's son for his father's causing the death of a house-owner's son by building the house badly; the death of a man's daughter because her father caused the death of another man's daughter.

The contracts naturally do not concern such criminal cases as the above, as a rule, but marriage contracts do specify death by strangling, drowning, precipitation from a tower or pinnacle of the temple, or by the iron sword, for a wife's repudiation of her husband. We are quite without evidence as to the executioner in all these cases.

Exile was inflicted for incest with a daughter; disinheritance for incest with a stepmother, or for repeated unfilial conduct. Sixty strokes of an ox-hide scourge were awarded for a brutal assault on a superior, both being amelu. Branding (perhaps the equivalent of degradation to slavery) was the penalty for slander of a married woman or vestal. Permanent deprivation of office fell upon the corrupt judge. Enslavement befell the extravagant wife and unfilial children. Imprisonment was common, but is not recognized by the Code.

The commonest of all penalties was a fine. This is awarded by the Code for corporal injuries to a muskinu or to a slave (paid to his master); for damages done to property, or for breach of contract. The restoration of goods appropriated, illegally bought, or damaged by neglect, was usually accompanied by a fine, giving it the form of multiple restoration. This might be double, treble, fourfold, fivefold, sixfold, tenfold, twelvefold, even thirtyfold, according to the enormity of the offence.

The Code recognized the importance of intent. A man who killed another in a quarrel must swear he did not do so intentionally, and was then only fined according to the rank of the deceased. The Code does not say what would be the penalty of murder, but death is so often awarded where death is caused, that we can hardly doubt that the murderer was put to death. If the assault only led to injury and was unintentional, the assailant in a quarrel had to pay the doctor's fees. A brander, induced to remove a slave's identification mark, could swear to his ignorance and was free. The owner of an ox that gored a man on the street was only responsible for damages if the ox was known by him to be vicious - even if it caused death. If the mancipium died a natural death under the creditor's hand, the creditor was scot-free. In ordinary cases, responsibility was not demanded for accident or for more than proper care. Poverty excused bigamy on the part of a deserted wife.

On the other hand, carelessness and neglect were severely punished, as in the case of the unskillful physician, if it led to loss of life or limb, his hands were cut off; a slave had to be replaced, the loss of his eye paid for to half his value; a veterinary surgeon who caused the death of an ox or donkey paid quarter value; a builder, whose careless workmanship caused death, lost his life or paid for it by the death of his child, replaced slave or goods, and in any case had to rebuild the house, or make good any damages due to defective building and repair the defect as well. The boat-builder had to make good any defect of construction or damage due to it for a year's warranty.

Throughout the Code, respect is paid to status. Suspicion was not enough. The criminal must be taken in the act, e.g. the adulterer, etc. A man could not be convicted of theft unless the goods were found in his possession.

In the case of a lawsuit, the plaintiff preferred his own plea. There is no trace of professional advocates, but the plea had to be in writing, and the notary doubtless assisted in the drafting of it. The judge saw the plea, called the other parties before him, and sent for the witnesses. If these were not at hand, he might adjourn the case for their subpoena, specifying a time up to six months. Pledges might be entered into to produce the witnesses on a fixed day.

The more important cases, especially those involving life and death, were tried by a bench of judges. With the judges were associated a body of elders who shared in the decision, but whose exact function is not yet clear. Agreements, declarations and non-contentious cases were usually witnessed by one judge and twelve elders.

Parties and witnesses were put on oath. The penalty for false witness was usually what would have been awarded the convicted criminal. In matters beyond the knowledge of men, as the guilt or innocence of an alleged practitioner of magic or a suspected wife, the ordeal by water was used. The accused jumped into the sacred river, and the innocent swam while the guilty drowned. The accused could clear himself by oath where his own knowledge was alone available.

The plaintiff could swear to his loss by brigands, as to goods claimed, the price paid for a slave purchased abroad, or the sum due to him. But great stress was laid on the production of written evidence. It was a serious thing to lose a document. The judges might be satisfied of its existence and terms by the affidavit of the witnesses to it, and then issue an order that whenever found it should be given up. Contracts annulled were ordered to be broken. The court might even go a journey to view the property and take with them the sacred symbols whereby oath was made.

The decision given was embodied in writing, sealed and witnessed by the judges, the elders, witnesses, and a scribe. Women might act in all these capacities. The parties swore an oath, embodied in the document, to observe its stipulations. Each took a copy, and one was held by the scribe to be stored in the archives.

Appeal to the king was allowed and is well attested. The judges at Babylon seem to have formed a superior court to those of provincial towns, but a defendant might elect to answer the charge before the local court, and refuse to plead at Babylon.

Finally, it may be noted that many immoral acts, such as the use of false weights, lying, etc., that could not be brought into court, are severely denounced in the Omen Tablets as likely to bring the offender into "the hand of God" as opposed to "the hand of the king."

## BABYLONIAN LITERATURE

The Babylonians were an ancient culture located in what is now Iraq. They had very advanced systems of writing, science and mathematics for their time. Most of what we have from the Babylonians was inscribed in cuneiform with a metal stylus on tablets of clay, called laterculae coctiles by Pliny the Elder; papyrus seems to have been also employed, but it has perished.

There were libraries in most towns and temples; an old Sumerian proverb averred that "he who would excel in the school of the scribes must rise with the dawn." Women as well as men learned to read and write, and in Semitic times, this involved a knowledge of the extinct Sumerian language, and a complicated and extensive syllabary.

A considerable amount of Babylonian literature was translated from Sumerian originals, and the language of religion and law long continued to be the old agglutinative language of Sumer. Vocabularies, grammars, and interlinear translations were compiled for the use of students, as well as commentaries on the older texts and explanations of obscure words and phrases. The characters of the syllabary were all arranged and named, and elaborate lists of them were drawn up.

Many Babylonian literary works have come down to us. One of the most famous of these was the Epic of Gilgamesh, in twelve books, translated from the original Sumerian by a certain Sin-liqi-unninni, and arranged upon an astronomical principle. Each division contains the story of a single adventure in the career of Gilgamesh. The whole story is a composite product, and it is probable that some of the stories are artificially attached to the central figure.

Another epic was that of the "Creation", whose object was to glorify Bel-Marduk by describing his contest with Tiamat, the dragon of chaos. In the first book, an account is given of the creation of the world from the primeval deep, and the birth of the gods of light.

Then comes the story of the struggle between the gods of light and the powers of darkness, and the final victory of Marduk, who clove Tiamat asunder, forming the heaven from half of her body and the earth from the other. Marduk next arranged the stars in order, along with the sun and moon, and gave them laws they were never to transgress. After this, the plants and animals were created, and finally man. Marduk here takes the place of Ea, who appears as the creator in the older legends, and is said to have fashioned man from clay.

The legend of Adapa, the first man - a portion of which was found in the record-office of the Egyptian king Akhenaton at Tell-el-Amarna - explains the origin of death. Adapa, while fishing, had broken the wings of the south wind, and was accordingly summoned before the tribunal of Anu in heaven. Ea counselled him not to eat or drink anything there. He followed this advice, and thus refused the food that would have made him and his descendants immortal.

Among the other legends of Babylonia may be mentioned those of Namtar, the plague-demon; of Erra, the pestilence; of Etanna and of Anzu. Hades, the abode of Ereshkigal or Allatu, had been entered by Nergal, who, angered by a message sent to her by the gods of the upper world, ordered Namtar to strike off her head. She, however, declared that she would submit to any conditions imposed on her, and would give Nergal the sovereignty of the earth. Nergal accordingly relented, and Allatu became the queen of the infernal world. Etanna conspired with the eagle to fly to the highest heaven.

The first gate, that of Anu, was successfully reached; but in ascending still farther to the gate of Ishtar, the strength of the eagle gave way, and Etanna was dashed to the ground. As for the storm-god Anzu, we are told that he stole the tablets of destiny, and therewith the prerogatives of Enlil. God after god was ordered to pursue him and recover them, but it would seem that it was only by a stratagem that they were finally regained.

Besides the purely literary works, there were others of varied nature, including collections of letters, partly official, partly private. Among them the most interesting are the letters of Hammurabi, which have been edited by Leonard William King.

Assyrian culture and literature came from Babylonia, but even here there was a difference between the two countries.

There was little in Assyrian literature that was original, and education, general in Babylonia, was mostly restricted to a single class in the northern kingdom. In Babylonia, it was of very old standing.

Under the second Assyrian empire, when Nineveh had become a great centre of trade, Aramaic - the language of commerce and diplomacy - was added to the number of subjects that the educated class was required to learn.

Under the Seleucids, Greek was introduced into Babylon, and fragments of tablets have been found with Sumerian and Assyrian (i.e. Semitic Babylonian) words transcribed into Greek letters.

**Babylonian Mathematics**

Babylonian clay tablet YBC 7289 with annotations. The diagonal displays an approximation of the square root of 2 in four sexagesimal figures, which is about six decimal figures.

1 + 24/60 + 51/602 + 10/603 = 1.41421296...

(Image by Bill Casselman)

Babylonian mathematics refers to any mathematics of the peoples of Mesopotamia (Iraq), from the days of the early Sumerians to the fall of Babylon in 539 BC. In contrast to the sparsity of sources in Egyptian mathematics, our knowledge of Babylonian mathematics is derived from some 400 clay tablets unearthed since the 1850s. Written in Cuneiform script, tablets were inscribed whilst the clay was moist, and baked hard in an oven or by the heat of the sun. The majority of recovered clay tablets date from 1800 to 1600 BC, and cover topics which include fractions, algebra, quadratic and cubic equations, the Pythagorean theorem, and the calculation of Pythagorean triples and possibly trigonometric functions (see Plimpton 322). The Babylonian tablet YBC 7289 gives an approximation to accurate to nearly six decimal places.

## Babylonian Numerals

The Babylonian system of mathematics was sexagesimal (base-60) numeral system. From this we derive the modern day usage of 60 seconds in a minute, 60 minutes in an hour, and 360 (60×6) degrees in a circle. The Babylonians were able to make great advances in mathematics for two reasons. Firstly, the number 60 is a Highly composite number, having divisors 1, 2, 3, 4, 5, 6, 10, 12, 15, 20, and 30, facilitating calculations with fractions. Additionally, unlike the Egyptians, Romans, the Babylonians and Indians had a true place-value system, where digits written in the left column represented larger values (much as in our base ten system: 734 = 7×100 + 3×10 + 4×1).

## Sumerian Mathematics (3000-2300 BC)

The earliest evidence of written mathematics dates back to the ancient Sumerians, who built the earliest civilization in Mesopotamia. They developed a complex system of metrology from 3000 BC. From 2600 BC onwards, the Sumerians wrote multiplication tables on clay tablets and dealt with geometrical exercises and division problems. The earliest traces of the Babylonian numerals also date back to this period.

## Old Babylonian Mathematics (2000-1600 BC)

The Old Babylonian period is the period to which most of the clay tablets on Babylonian mathematics belong, which is why the mathematics of Mesopotamia is commonly known as Babylonian mathematics. Some clay tablets contain mathematical lists and tables, others contain problems and worked solutions.

## Arithmetic

The Babylonians made extensive use of pre-calculated tables to assist with arithmetic. For example, two tablets found at Senkerah on the Euphrates in 1854, dating from 2000 BC, give lists of the squares of numbers up to 59 and the cubes of numbers up to 32. The Babylonians used the lists of squares together with the formulas

$$ab = \frac{(a+b)^2 - a^2 - b^2}{2}$$

$$ab = \frac{(a+b)^2 - (a-b)^2}{4}$$

to simplify multiplication.

The Babylonians did not have an algorithm for long division. Instead they based their method on the fact that:

$$\frac{a}{b} - a \times \frac{1}{b}$$

together with a table of reciprocals. Numbers whose only prime factors are 2, 3 or 5 have finite reciprocals in sexagesimal notation, and tables with extensive lists of these reciprocals have been found.

Reciprocals such as 1/7, 1/11, 1/13, etc. do not have finite representations in sexagesimal notation. To compute 1/13 or to divide a number by 13 the Babylonians would use an approximation such as:

$$\frac{1}{3}=\frac{7}{91}=7\times\frac{1}{91}\approx7\times\frac{1}{90}=7\times\frac{40}{3600}$$

**Algebra**

As well as arithmetical calculations, Babylonian mathematicians also developed algebraic methods of solving equations. Once again, these were based on pre-calculated tables.

To solve a quadratic equation the Babylonians essentially used the standard quadratic formula. They considered quadratic equations of the form

$x^2 + bx = c$

where here b and c were not necessarily integers, but c was always positive. They knew that a solution to this form of equation is

$$x=-\frac{b}{2}+\sqrt{\left(\frac{b}{2}\right)^2+c}$$

and they would use their tables of squares in reverse to find square roots. They always used the positive root because this made sense when solving "real" problems. Problems of this type included finding the dimensions of a rectangle given its area and the amount by which the length exceeds the width.

Tables of values of $n^3+n^2$ were used to solve certain cubic equations. For example, consider the equation

$ax^3 + bx^2 + c$.

'"Multiplying the equation by'" $a^2$ and dividing by $b^3$ gives

$$\left(\frac{ax}{b}\right)^3+\left(\frac{ax}{b}\right)^2=\frac{ca^2}{b^3}.$$

Substituting $y = ax/b$ gives

$$y^3+y^2=\frac{ca^2}{b^3}$$

which could now be solved by looking up the $n^3+n^2$ table to find the value closest to the right hand side. The Babylonians accomplished this without algebraic notation, showing a remarkable depth of understanding. However, they did not have a method for solving the general cubic equation.

## Geometry

The Babylonians may have known the general rules for measuring areas and volumes. They measured the circumference of a circle as three times the diameter and the area as one-twelfth the square of the circumference, which would be correct if p is estimated as 3. The volume of a cylinder was taken as the product of the base and the height, however, the volume of the frustum of a cone or a square pyramid was incorrectly taken as the product of the height and half the sum of the bases. The Pythagorean theorem was also known to the Babylonians. Also, there was a recent discovery in which a tablet used p as 3 and 1/8. The Babylonians are also known for the Babylonian mile, which was a measure of distance equal to about seven miles today. This measurement for distances eventually was converted to a time-mile used for measuring the travel of the Sun, therefore, representing time.

## Trigonometry

There is also evidence that the Babylonians first used trigonometric functions, based on a table of numbers written on the Babylonian cuneiform tablet, Plimpton 322 (circa 1900 BC), which can be interpreted as a table of secants.

## Plimpton 322

In each row of the Plimpton 322 tablet, the number in the second column can be interpreted as the shortest side s of a right triangle, and the number in the third column can be interpreted as the hypotenuse d of the triangle. The number in the first column is either the fraction $\frac{s^2}{l^2}$ or $\frac{d^2}{l^2}$, where l denotes the longest side of the same right triangle.

However, scholars differ on how these numbers were generated and why the Babylonians would have been interested in such tables.

Neugebauer (1951) argued for a number-theoretic interpretation, pointing out that this table provides a list of (pairs of numbers from) Pythagorean triples. For instance, line 11 of the table can be interpreted as describing a triangle with short side 3/4 and hypotenuse 5/4, forming the side:hypotenuse ratio of the familiar (3,4,5) right triangle. If p and q are two coprime numbers, then $\left(p^2 - q^2, 2pq, p^2 + q^2\right)$ form a Pythagorean triple, and all Pythagorean triples can be formed in this way. For instance, line 11 can be generated by this formula with p = 1 and q = 1/2. As Neugebauer argues, each line of the tablet can be generated by a pair (p,q) that are both regular numbers, integer divisors of a power of 60. This property of p and q being regular leads to a denominator that is regular, and therefore to a finite sexagesimal representation for the fraction in the first column. Neugebauer's explanation is the one followed e.g. by Conway and Guy (1996). However, as Robson points out, Neugebauer's theory fails to explain how the values of p and q were chosen: there are 92 pairs of coprime regular numbers up to 60, and only 15 entries in the table. In addition, it does not explain why the table entries are in the order they are listed in, nor what the numbers in the first column were used for.

Joyce (1995) provides a trigonometric explanation: the values of the first column can be interpreted as the squared cosine or tangent (depending on the missing digit) of the angle opposite the short side of the right triangle described by each row, and the rows are sorted by these angles in roughly one-degree increments. However, Robson argues on linguistic grounds that this theory is "conceptually anachronistic": it depends on too many other ideas not present in the record of Babylonian mathematics from that time.

Robson (2001,2002), based on prior work by Bruins (1949,1955) and others, instead takes an approach that in modern terms would be characterized as algebraic, though she describes it in concrete geometric terms and argues that the Babylonians would also have interpreted this approach geometrically. Robson bases her interpretation on another tablet, YBC 6967, from roughly the same time and place. This tablet describes a method for solving what we would nowadays describe as quadratic equations of the form $x - \frac{1}{x} = c$, by steps (described in geometric terms) in which the solver calculates a sequence of intermediate values $v_1 = c/2$, $v_2 = v_1^2$, $v_3 = 1 + v_2$, and $v_4 = v_3^{1/2}$, from which one can calculate $x = v_4 + v_1$ and $1/x = v_4 - v_1$. Robson argues that the columns of Plimpton 322 can be interpreted as the following values, for regular number values of x and 1/x in numerical order: $v_3$ in the first column, $v_1 = (x - 1/x)/2$ in the second column, and $v4 = (x + 1/x)/2$ in the third column. In this interpretation, x and 1/x would have appeared on the tablet in the broken-off portion to the left of the first column. For instance, row 11 of Plimpton 322 can be generated in this way for x = 2. Thus, the tablet can be interpreted as giving a sequence of worked-out exercises of the type solved by the method from tablet YBC 6967. It could, Robson suggests, have been used by a teacher as a problem set to assign to students.

## Neo-Babylonian Mathematics (626-539 BC)

The Neo-Babylonian Empire flourished during the Chaldean period of Mesopotamia, which marked the second flowering of Babylon as a capital city and center of study. This period provides the second source of Babylonian mathematics, though somewhat more vague than the Old Babylonian mathematics.

Since the rediscovery of the Babylonian civilization, it has become apparent that Greek and Hellenistic mathematicians and astronomers, and in particular Hipparchus, borrowed a lot from the Chaldeans.

Franz Xaver Kugler demonstrated in his book Die Babylonische Mondrechnung ("The Babylonian lunar computation", Freiburg im Breisgau, 1900) the following: Ptolemy had stated in his Almagest IV.2 that Hipparchus improved the values for the Moon's periods known to him from "even more ancient astronomers" by comparing eclipse observations made earlier by "the Chaldeans", and by himself. However Kugler found that the periods that Ptolemy attributes to Hipparchus had already been used in Babylonian ephemerides, specifically the collection of texts nowadays called "System B" (sometimes attributed to Kidinnu). Apparently

Hipparchus only confirmed the validity of the periods he learned from the Chaldeans by his newer observations.

It is clear that Hipparchus (and Ptolemy after him) had an essentially complete list of eclipse observations covering many centuries. Most likely these had been compiled from the "diary" tablets: these are clay tablets recording all relevant observations that the Chaldeans routinely made. Preserved examples date from 652 BC to AD 130, but probably the records went back as far as the reign of the Babylonian king Nabonassar: Ptolemy starts his chronology with the first day in the Egyptian calendar of the first year of Nabonassar, i.e., 26 February 747 BC.

This raw material by itself must have been hard to use, and no doubt the Chaldeans themselves compiled extracts of e.g., all observed eclipses (some tablets with a list of all eclipses in a period of time covering a saros have been found). This allowed them to recognise periodic recurrences of events. Among others they used in System B (cf. Almagest IV.2):

- 223 (synodic) months = 239 returns in anomaly (anomalistic month) = 242 returns in latitude (draconic month). This is now known as the saros period which is very useful for predicting eclipses.
- 251 (synodic) months = 269 returns in anomaly
- 5458 (synodic) months = 5923 returns in latitude
- 1 synodic month = 29;31:50:08:20 days (sexagesimal; 29.53059413... days in decimals = 29 days 12 hours 44 min $3\frac{1}{3}$ s)

The Babylonians expressed all periods in synodic months, probably because they used a lunisolar calendar. Various relations with yearly phenomena led to different values for the length of the year.

Similarly various relations between the periods of the planets were known. The relations that Ptolemy attributes to Hipparchus in Almagest IX.3 had all already been used in predictions found on Babylonian clay tablets.

All this knowledge was transferred to the Greeks probably shortly after the conquest by Alexander the Great (331 BC).

According to the late classical philosopher Simplicius (early 6th century AD), Alexander ordered the translation of the historical astronomical records under supervision of his chronicler Callisthenes of Olynthus, who sent it to his uncle Aristotle. It is worth mentioning here that although Simplicius is a very late source, his account may be reliable. He spent some time in exile at the Sassanid (Persian) court, and may have accessed sources otherwise lost in the West.

It is striking that he mentions the title tèresis (Greek: guard) which is an odd name for a historical work, but is in fact an adequate translation of the Babylonian title massartu meaning "guarding" but also "observing". Anyway, Aristotle's pupil Callippus of Cyzicus introduced his 76-year cycle, which improved upon the 19-year Metonic cycle, about that time. He had the first year of his first cycle start at the summer solstice of 28 June 330 BC (Julian proleptic date), but later he seems to have counted lunar months from the first month after Alexander's decisive battle at Gaugamela in fall 331 BC. So Callippus may have obtained his data from Babylonian sources and his calendar may have been anticipated by Kidinnu.

Also it is known that the Babylonian priest known as Berossus wrote around 281 BC a book in Greek on the (rather mythological) history of Babylonia, the Babyloniaca, for the new ruler Antiochus I; it is said that later he founded a school of astrology on the Greek island of Kos. Another candidate for teaching the Greeks about Babylonian astronomy/astrology was Sudines who was at the court of Attalus I Soter late in the 3rd century BC.

In any case, the translation of the astronomical records required profound knowledge of the cuneiform script, the language, and the procedures, so it seems likely that it was done by some unidentified Chaldeans. Now, the Babylonians dated their observations in their lunisolar calendar, in which months and years have varying lengths (29 or 30 days; 12 or 13 months respectively). At the time they did not use a regular calendar (such as based on the Metonic cycle like they did later), but started a new month based on observations of the New Moon. This made it very tedious to compute the time interval between events.

What Hipparchus may have done is transform these records to the Egyptian calendar, which uses a fixed year of always 365 days (consisting of 12 months of 30 days and 5 extra days): this makes computing time intervals much easier. Ptolemy dated all observations in this calendar. He also writes that "All that he (=Hipparchus) did was to make a compilation of the planetary observations arranged in a more useful way" (Almagest IX.2). Pliny states (Naturalis Historia II.IX(53)) on eclipse predictions: "After their time (=Thales) the courses of both stars (=Sun and Moon) for 600 years were prophecied by Hipparchus, ...". This seems to imply that Hipparchus predicted eclipses for a period of 600 years, but considering the enormous amount of computation required, this is very unlikely. Rather, Hipparchus would have made a list of all eclipses from Nabonasser's time to his own.

Other traces of Babylonian practice in Hipparchus' work are:

- first Greek known to divide the circle in 360 degrees of 60 arc minutes.
- first consistent use of the sexagesimal number system.
- the use of the unit pechus ("cubit") of about 2° or 2½°.
- use of a short period of 248 days = 9 anomalistic months.

**PERSIAN EMPIRE**

| Empires of Persia · Kings of Persia | BCE |
|---|---|
| Zayandeh Rud civilization | 5th millenium |
| Jiroft civilization (Aratta?) | 3000-5th cent. |
| Proto-Elamite civilization | 3500-2800 |
| Elamite dynasties | 2800-550 |
| Kingdom of Mannai | 10th-7th cent. |
| Median Empire | 728-550 |
| Achaemenid Empire | 550-330 |
| Seleucid Empire | 330-150 |
| Parthian Empire | 248-CE 224 CE |
| Sassanid Empire | 224-651 |

| | |
|---|---|
| Patriarchal Caliphate | 637-651 |
| Umayyad Caliphate | 661-750 |
| Abbasid Caliphate | 750-1258 |
| Tahirid dynasty | 821-873 |
| Alavid dynasty | 864-928 |
| Sajid dynasty | 889/890-929 |
| Saffarid dynasty | 861-1003 |
| Samanid dynasty | 875-999 |
| Ziyarid dynasty | 928-1043 |
| Buwayhid dynasty | 934-1055 |
| Ghaznavid Empire | 963-1187 |
| Ghori dynasty | 1149-1212 |
| Seljukid Empire | 1037-1194 |
| Khwarezmid dynasty | 1077-1231 |
| Ilkhanate | 1256-1353 |
| Muzaffarid dynasty | 1314-1393 |
| Chupanid dyansty | 1337-1357 |
| Jalayerid dynasty | 1339-1432 |
| Timurid Empire | 1370-1506 |
| Qara Qoyunlu Turcomans | 1407-1468 |
| Aq Qoyunlu Turcomans | 1378-1508 |
| Safavid dynasty | 1501-1722* |
| Hotaki Ghilzai dynasty | 1722-1729 |
| Afsharid dynasty | 1736-1802 |
| Zand dynasty | 1750-1794 |
| Qajar dynasty | 1781-1925 |
| Pahlavi dynasty | 1925-1979 |
| Islamic Revolution | 1979 |
| Provisional Government | 1979-1980 |
| Islamic Republic of Iran | since 1980 |

* or 1736

The Persian Empire was a series of historical empires that ruled over the Iranian plateau, the old Persian homeland, and beyond in Western Asia, Central Asia and the Caucasus. The Persian empire has become what we know today as Persia or more commonly, Iran.

The most widespread entity considered to have been a Persian Empire was the Achaemenid Empire (550-330 BC)- famous in antiquity as the foe of the classical Greek states- a united Iranian kingdom that originated in the region now known as Pars province of Iran.

It was formed under Cyrus the Great, who overthrew the empire of the Medes, and conquered much of the Middle East, including the territories of the Babylonians, the Phoenicians, and the Lydians. Cyrus' son, Cambyses, continued Cyrus' conquests by conquering Egypt.

Most of the successive states in Greater Iran prior to March 1935 are collectively called the Persian Empire by Western historians.

Virtually all the successor empires of Persia were major regional and some major international powers in their day.

**Naming Dispute**

Persia has long been used by the West to describe the nation of Iran, its people, and its ancient empires. It derives from the ancient Greek name for Iran's maritime province, called Fars in the modern Persian language, Pars in Middle Persian and Parsa in Old Persian, a word meaning "above reproach". Persis is the Hellenized form of Pars, and through the Latinized word Persia, the other European nations came to use this word for the region. This area was the core of the original Persian Empire.

Most foreigners referred to the entire country as Persia until March 21, 1935, when Reza Shah Pahlavi formally asked the international community to call the country Iran-a name that the people of Persia, themselves, used to refer to their country. An anology would be demanding Germany be called Deutschland.

## HISTORY

### *Achaemenid Empire (550 BC-330 BC)*

The earliest known record of the Persians comes from an Assyrian inscription from c. 844 BC that calls them the Parsu (Parsuash, Parsumash) and mentions them in the region of Lake Urmia alongside another group, the Madai (Medes). For the next two centuries, the Persians and Medes were at times tributary to the Assyrians. The region of Parsuash was annexed by Sargon of Assyria around 719 BC. Eventually the Medes came to rule an independent Median Empire, and the Persians were subject to them.

The Achaemenids were the first to create a centralized state in Persia, founded by Achaemenes (Hakhamanish), chieftain of the Persians around 700 BC.

Around 653 BC, the Medes came under the domination of the Scythians, and Teispes, the son of Achaemenes, seems to have led the nomadic Persians to settle in southern Iran around this time - eventually establishing the first organized Persian state in the important region of Anshan as the Elamite kingdom was permanently destroyed by the Assyrian ruler Ashurbanipal (640 BC). The kingdom of Anshan and its successors continued to use Elamite as an official language for quite some time after this, although the new dynasts spoke Persian, an Indo-Iranian tongue.

Teispes' descendants may have branched off into two lines, one line ruling in Anshan, while the other ruled the rest of Persia. Cyrus II the Great united the separate kingdoms around 559 BC. At this time, the Persians were still tributary to the Median Empire ruled by Astyages. Cyrus rallied the Persians together, and in 550 BC defeated the forces of Astyages, who was then captured by his own nobles and turned over to the triumphant Cyrus, now Shah of a unified Persian kingdom. As Persia assumed control over the rest of Media and their large empire, Cyrus led the united Medes and Persians to still more conquest. He took Lydia in Asia Minor, and carried his arms eastward into central Asia. Finally in 539 BC, Cyrus marched triumphantly into the ancient city of Babylon. After this victory, he set the standards of a benevolent conqueror by issuing the

Cyrus Cylinder, the first charter of human rights. Cyrus was killed in 530 BC during a battle against the Massagetae or Sakas.

Cyrus' son, Cambyses II, annexed Egypt to the Achaemenid Empire. The empire then reached its greatest extent under Darius I. He led conquering armies into the Indus River valley and into Thrace in Europe. A punitive raid against Greece was halted at the Battle of Marathon. His son Xerxes I tried to subdue the Greeks, but his army was defeated at the Battle of Plataea 479 BC.

The Achaemenid Empire was the largest and most powerful empire the world had yet seen. More importantly, it was well managed and organized. Darius divided his realm into about twenty satrapies (provinces) supervised by satraps, or governors, many of whom had personal ties to the Shah. He instituted a systematic tribute to tax each province. He took the advanced postal system of the Assyrians and expanded it. Also taken from the Assyrians was the usage of secret agents of the king, known as the King's Eyes and Ears, keeping him informed.

Darius improved the famous Royal Road and other ancient trade routes, thereby connecting far reaches of the empire. He may have moved the administration center from Fars itself to Susa, near Babylon and closer to the center of the realm. The Persians allowed local cultures to survive, following the precedent set by Cyrus the Great. This was not only good for the empire's subjects, but ultimately benefited the Achaemenids, since the conquered peoples felt no need to revolt.

It may have been during the Achaemenid period that Zoroastrianism reached South-Western Iran, where it came to be accepted by the rulers and through them became a defining element of Persian culture. The religion was not only accompanied by a formalization of the concepts and divinities of the traditional (Indo-)Iranian pantheon but also introduced several novel ideas, including that of free will, which is arguably Zoroaster's greatest contribution to religious philosophy. Under the patronage of the Achaemenid kings, and later as the de-facto religion of the state, Zoroastrianism would reach all corners of the empire. In turn, Zoroastrianism would be subject to the first sycretic influences, in particular from the Semitic lands to the west, from which the divinities of the religion would gain astral and planetary aspects and from where the temple cult originates. It was also during the Achaemenid era that the sacerdotal Magi would exert their influence on the religion, introducing many of the practices that are today identified as typically Zoroastrian, but also introducing doctrinal modifications that are today considered to be revocations of the original teachings of the prophet.

The Achaemenid Empire united people and kingdoms from every major civilization in south west Asia. For the first time in history, people from very different cultures were in contact with one another under one ruler.

### *Hellenistic Persia (330 BC-250 BC )*

The Achaemenid dynasty never managed to realise the conquest of the Greeks, but kept their threat at a minor level, often by encouraging their internal fighting. However, the Achaemenid Empire's weakness was exposed to the Greeks in 401 BC, when a rebel prince, Cyrus the Younger, hired 14,000 Greek mercenaries to help secure his claim to the imperial throne (see Xenophon, Anabasis). This demonstrated the military problems of the Achaemenid forces when dealing with an effective phalanx of spearmen.

Philip II of Macedon managed to unify most of Greece under his control, and decided to take advantage of Achaemenid weakness when, after the death of Artaxerxes III Ochus in 338, the Persian Empire had no strong leader. On Philip's death in 336, his son and heir continued the attack on the Empire. He turned out to be one of the most effective generals in history. The Achaemenid monarch, Darius III was an aged man with a reputation for bravery gained in his youth, but in the event was no match for Alexander. The greatest empire of the time collapsed in only eight years.

Alexander landed in Asia Minor in 334 BC. His armies quickly swept through Lydia, Phoenicia, and Egypt, before defeating Darius III at Gaugamela (331) and capturing the capital at Susa. The last Achaemenid resistance was at the "Persian Gates" between Susa and near the royal palace at Persepolis. The Achaemenid Empire was now in Alexander's hands. Along his route of conquest, Alexander founded many colony cities, often named "Alexandria". For the next several centuries, these cities served to greatly extend Greek, or Hellenistic, culture in Persia.

Alexander's empire broke up shortly after his death, and Alexander's general, Seleucus I Nicator, tried to take control of Persia, Mesopotamia, and later Syria and Asia Minor. His ruling family is known as the Seleucid Dynasty. However he was killed in 281 BC by Ptolemy Keraunos before he could conquer Greece and Macedonia.

Greek colonization continued until around 250 BC; Greek language, philosophy, and art came with the colonists. Throughout Alexander's former empire, Greek became the common tongue of diplomacy and literature. Trade with China had begun in Achaemenid times along the so-called Silk Road; but during the Hellenistic period it began in earnest. The overland trade brought about some fascinating cultural exchanges. Buddhism came in from India, while Zoroastrianism traveled west to influence Judaism. Incredible statues of the Buddha in classical Greek styles have been found in Persia and Afghanistan, illustrating the mix of cultures that occurred around this time (See Greco-Buddhism), although it is possible that Greco-Buddhist art dates from Achaemenid times when Greek artists worked for the Persians.

Although recently discovered cuneiform evidence (e.g., the Babylonian Chronicles from the Hellenistic Period) show how much continuity there was in the Eastern civilization, it can not be denied that the Seleucid kingdom began to decline after about a century. The eastern provinces of Bactria and Parthia broke off in 238 BC.

King Antiochus III's military leadership kept Parthia from overrunning Persia itself, but when he tried to intervene in Greece, his successes alarmed the burgeoning Roman Republic. Roman legions began to attack the kingdom.

At the same time, the Seleucids had to contend with the revolt of the Maccabees in Judea and the expansion of the Kushan Empire to the east. The empire fell apart and was conquered by Parthia and Rome.

### *Parthian Empire (250 BC-AD 226)*

Its rulers, the Arsacid dynasty, belonged to an Iranian tribe that had settled there during the time of Alexander. They declared their independence from the Seleucids in 238 BC, but their attempts to unify Iran were thwarted until after Mithridates I advent to the Parthian throne in about 170 BC.

The Parthian Confederacy shared a border with Rome along the upper Euphrates River. The two polities became major rivals, especially over control of Armenia. Heavily-armoured Parthian cavalry (cataphracts) supported by mounted archers proved a match for Roman legions, as in the Battle of Carrhae in which the Parthian General Surena defeated Marcus Licinius Crassus of Rome. Wars were very frequent, with Mesopotamia serving as the battleground. The family of the Persian Empire now goes by the name of Rahbar.

During the Parthian period, Hellenistic customs partially gave way to a resurgence of Persian culture. However, the empire lacked political unity.

The administration was shared between Seven Parthian clans who constituted the Dahae Confederation, each of these clans governed a province of the empire. Suren-Pahlav Clan, Karen-Pahlav Clan and Mihran Clan were the most influential ones. By the 1st century BC, Parthia was decentralized, ruled by feudal nobles. Wars with Rome to the west and the Kushan Empire to the northeast drained the country's resources.

Parthia, now impoverished and without any hope to recover the lost territories, was demoralized. The kings had to give more concessions to the nobility, and the vassal kings sometimes refused to obey. Parthia's last ruler Artabanus IV had an initial success in putting together the crumbling state.

However, the fate of the Arsacid Dynasty was doomed when in AD 224, the Persian vassal king Ardashir revolted. Two years later, he took Ctesiphon, and this time, it meant the end of Parthia. It also meant the beginning of the third Persian Empire, ruled by the Sassanid kings. Sassanids were from the province of Persis, native to the first Persian Empire, the Achaemenids.

### *Sassanid Empire (226-651)*

Ardashir I, led a rebellion against the Parthian Confederacy in an attempt to revive the glory of the previous empire and to legitimize the hellenized form of Zoroastrianism practised in south western Iran. In two years he was the Shah of a new Persian Empire.

The Sassanid dynasty (also Sassanian) (named for Ardashir's grandfather) was the first dynasty native to the Pars province since the Achaemenids; thus they saw themselves as the successors of Darius and Cyrus. They pursued an aggressive expansionist policy. They recovered much of the eastern lands that the Kushans had taken in the Parthian period. The Sassanids continued to make war against Rome; a Persian army even captured the Roman Emperor Valerian in 260.

The Sassanid Empire, unlike Parthia, was a highly centralized state. The people were rigidly organized into a caste system: Priests, Soldiers, Scribes, and Commoners. Zoroastrianism was finally made the official state religion, and spread outside Persia proper and out into the provinces. There was sporadic persecution of other religions. The Eastern Orthodox Church was particularly persecuted, but this was in part due to its ties to the Roman Empire. The Nestorian Christian church was tolerated and sometimes even favored by the Sassanids.

The wars and religious control that had fueled The Sassanid empire's early successes eventually contributed to its decline. The eastern regions were conquered by the White Huns in the late 5th century. Adherents of a radical religious sect, the Mazdakites, revolted around the same time. Khosrau I was able to recover his empire and expand into the Christian countries of Antioch and Yemen. Between 605 and 629, Sassanids successfully annexed Levant and Roman Egypt and pushed into Anatolia.

However, a subsequent war with the Romans utterly destroyed the empire. In the course of the protracted conflict, Sassinid armies reached Constantinople, but could not defeat the Byzantines there. Meanwhile, the Byzantine Emperor Heraclius had successfully outflanked the Persian armies in

Asia Minor and attacked the empire from the rear while the main Iranian army along with its top Eran Spahbods were far from battlefields. This resulted in a crushing defeat for Sassanids in Northern Mesopotamia. The Sassanids had to give up all their conquered lands and retreat. This defeat was mentioned in Qur'an as a "victory for believers," referring to the Romans, who were monotheists, in contrast to the pagan Sassinids. (Note: The official religion of the Sassanid empire was Zoroastrianism. It is not an Abrahamic/Semitic religion like Christianity or Islam, so it would be classified as "Pagan" by the followers of those religions even though it was monotheistic).

Following the advent of Islam and collapse of Sassanid Empire, Persians came under the subjection of Arab rulers for almost two centuries before native Persian dynasties could gradually drive them out. In this period a number of small and numerically inferior Arab tribes migrated to inland Iran.

Also some Turkic tribes settled in Persia between the 9th and 12th centuries.

In time these peoples were integrated into numerous Persian populations and adopted Persian culture and language while Persians retained their culture with minimal influence from outside.

## Fall of Persian Empire

Some of the information in this article or section may not be verified by reliable sources. It should be checked for inaccuracies and modified to cite reliable sources.

The explosive growth of the Arab Caliphate coincided with the chaos caused by the defeat of Sassanids in wars with the Byzantine Empire. Most of the country was conquered between 643 to 650.

Yazdgerd III, the last Sasanian emperor, died ten years after he lost his empire to the newly-formed Muslim Caliphate. He tried to recover some of what he lost with the help of the Turks, but they were easily defeated by Muslim armies. Then he sought the aid of the Chinese Tang dynasty. However, the Chinese help did not avail and Arab muslims ultimately defeated the Chinese forces in the battle of Talas, a century after

Yazdgerd's death. The Umayyads would rule Persia for a hundred years. The Arab conquest dramatically changed life in Persia. Arabic became the new lingua franca, Islam eventually replaced Zoroastrianism, and mosques were built. A new language, religion, and culture were added to the Iranian cultural milieu.

In 750 the Umayyads were ousted from power by the Abbasid dynasty. By that time, Persians had come to play an important role in the bureaucracy of the empire. The caliph Al-Ma'mun, whose mother was Persian, moved his capital away from Arab lands into Merv in eastern Iran. In 819, Samanids carved out an semi-independent state in eastern Persia to become the first native rulers after the Arabic conquest. They made Samarqand, Bukhara and Herat their capitals and revived the Persian language and culture. Zoroastrian clerics complied and authored major religious texts, such as the Denkard, in Pahlavi. It was approximately during this age, when the poet Firdawsi finished the Shahnameh, an epic poem retelling the history of the Iranian kings. This epic was completed by 1008 AD.

In 913, western Persia was conquered by the Buwayhid, a Deylamite tribal confederation from the shores of the Caspian Sea. They made the city of Shiraz their capital. The Buwayhids destroyed Islam's former territorial unity. Rather than a province of a united Muslim empire, Iran became one nation in an increasingly diverse and cultured Islamic world.

## Turkic Rule (1037-1219)

The Muslim world was shaken again in 1037 with the invasion of the Seljuk Turks from the northeast. The Seljuks created a very large Middle Eastern empire. The Seljuks built the fabulous Friday Mosque in the city of Isfahan. The famous Persian mathematician and poet, Omar Khayyám, wrote his Rubaiyat during Seljuk times. In the early 13th century the Seljuks lost control of Persia to another group of Turks from Khwarezmia, near the Aral Sea. The Shahs of the Khwarezmid Empire later ruled.

## Mongols and their Successors (1219-1500)

In 1218, Genghis Khan sent ambassadors and merchants to the city of Otrar, on the northeastern confines of the Khwarizm shahdom. The governor of Otrar had these envoys executed. Genghis attacked Otrar in 1219, Samarkand and other cities of the northeast. Genghis' grandson, Hulagu Khan, finished the invasions that Genghis had begun when he defeated Khwarzim Empire, Baghdad, and much of the rest of the Middle East from 1255 to 1258. Persia temporarily became the Ilkhanate, a division of the vast Mongol Empire.

In 1295, after Ilkhan Mahmud Ghazan converted to Islam, he renounced all allegiance to the Emperor Chengzong of Yuan China who had recently succeeded his grandfather Kublai Khan as Great Khan. The Ilkhans patronized the arts and learning in the fine tradition of Iranian Islam; indeed, they helped to repair much of the damage of the Mongol conquests.

In 1335, the death of Abu Sa'id, the last well-recognized Ilkhan, spelled the end of the Ilkhanate. Though Arpa Ke'un was declared Ilkhan his authority was disputed and the Ilkhanate was splintered into a number of small states. This left Persia vulnerable to conquest at the hands of Timur the Lame or Tamerlane, a Central Asian conqueror seeking to revive the Mongol Empire. He ordered the attack of Persia beginning around 1370 and robbed the region until his death in 1405. Timur was an even greater murderer than Genghis had been. In Isfahan, for instance, he was responsible for the murder of 70,000 people so that he could build towers with their skulls. He conquered a wide area and made his own city of Samarkand rich, but he made no effort to forge a lasting empire. The Persian Empire was essentially in ruins.

For the next hundred years Persia was not a unified state. It was ruled for a while by descendants of Timur, called the Timurid emirs. Toward the end of the 15th century, Persia was taken over by the Emirate of the White Sheep Turkmen (Ak Koyunlu). But there was little unity and none of the sophistication that had defined Iran during the glory days of Islam.

## Safavid Dynasty (1500-1722)

The Safavid Dynasty hailed from the town of Ardabil in the region of Azarbaijan. The Safavid Shah Ismail I overthrew the White Sheep (Akkoyunlu) Turkish rulers of Persia to found a new native Persian empire. Ismail expanded Persia to include all of present-day Azerbaijan, Iran, and Iraq, plus much of Afghanistan. Ismail's expansion was halted by the Ottoman Empire at the Battle of Chaldiran in 1514, and war with the Ottomans became a fact of life in Safavid Iran.

Safavid Persia was a violent and chaotic state for the next seventy years, but in 1588 Shah Abbas I of Safavid ascended to the throne and instituted a cultural and political renaissance. He moved his capital to Isfahan, which quickly became one of the most important cultural centers in the Islamic world. He made peace with the Ottomans. He reformed the army, drove the Uzbeks out of Iran and into modern-day Uzbekistan, and (with English help) recaptured the island of Hormuz from the Portuguese. Abdur Razzaq was the Persian ambassador to Calicut, India, and wrote vividly of his experiences there.

The Safavids were followers of Shi'a Islam, and under them Persia (Iran) became the largest Shi'a country in the Muslim world, a position Iran still holds today.

Under the Safavids Iran enjoyed its last period as a major imperial power. In 1639, a final border was agreed upon with the Ottoman Empire with the Treaty of Qasr-e Shirin; which delineates the border between the Republic of Turkey and Iran and also that of between Iraq and Iran, today.

## Persia and Europe (1722-1914)

Throughout the Middle Ages, the natural philosophy and mathematics of ancient Greeks were furthered and preserved within the Muslim world. During this period, Persia became a centre for the manufacture of scientific instruments, retaining its reputation for quality well into the 19th century.

In 1722, the Safavid state collapsed. That year saw the first European invasion of Persia since the time of Alexander: Peter the Great, Emperor of Imperial Russia, invaded from the northwest as part of a bid to dominate central Asia. Ottoman forces accompanied the Russians, successfully laying siege to Isfahan.

The country was able to weather the invasions; neither the Russians nor the Turks gained any territory. However, the

Safavids were severely weakened, and that same year (1722), the Afghans launched a bloody battle in response to the Safavids' attempts on trying to forcefully convert them from Sunni to Shi'a sect of Islam. The last Safavid shah was executed, and the dynasty came to an end.

The Persian empire experienced a temporary revival under Nader Shah in the 1730s and 1740s. Nadir drove out the Russians and confined the Afghans to their present home in Afghanistan. He launched many successful campaigns against Persia's old enemies, the nomadic khanates of Central Asia; most of them were destroyed or absorbed into Persia.

In 1739, he attacked and looted Delhi, the capital of Moghul India. However, his empire declined after his death. His rule was followed by the weak and short-lived Zand dynasty. Iran was left unprepared for the worldwide expansion of European colonial empires in the late 18th century and throughout the 19th century.

Persia found relative stability in the Qajar dynasty, ruling from 1779 to 1925, but lost hope to compete with the new industrial powers of Europe; Persia found itself sandwiched between the growing Russian Empire in Central Asia and the expanding British Empire in India. Each carved out pieces from the Persian empire that became Bahrain, Azerbaijan, Turkmenistan, Uzbekistan, and Afghanistan amongst other previous provinces.

Although Persia was never directly invaded, it gradually became economically dependent on Europe. The Anglo-Russian Convention of 1907 formalised Russian and British spheres of influence over the north and south of the country, respectively, where Britain and Russia each created a "sphere of influence", where the colonial power had the final "say" on economic matters.

At the same time Mozzafar-al-Din shah had granted a concession to William Knox D'Arcy, later the Anglo-Persian Oil Company, to explore and work the newly-discovered oil fields at Masjid Soleiman in southwest Persia, which started production in 1914. Winston Churchill, as First Sea Lord to the British Admiralty, oversaw the conversion of the Royal Navy to oil-fired battleships and partially nationalized it prior to the start of war. A small Anglo-Persian force was garrisoned there to protect the field from some hostile tribal factions.

## World War I and the Interbellum (1914-1935)

Persia was drawn into the periphery of World War I because of its strategic position between Afghanistan and the warring Ottoman, Russian, and British Empires. In 1914 Britain sent a military force to Mesopotamia to deny the Ottomans access to the Persian oilfields. The German Empire retaliated on behalf of its ally by spreading a rumour that Kaiser Wilhelm II of Germany had converted to Islam, and sent agents through Iran to attack the oil fields and raise a Jihad against British rule in India. Most of those German agents were captured by Persian, British and Russian troops who were sent to patrol the Afghan border, and the rebellion faded away. This was followed by a German attempt, assisted by his mainly Swedish bodyguard, to abduct and control Ahmad Shah Qajar, which was foiled at the last moment.

In 1916 the fighting between Russian and Ottoman forces to the north of the country had spilled down into Persia; Russia gained the advantage until most of her armies collapsed in the wake of the Russian Revolution of 1917. This left the Caucasus unprotected, and the Caucasian and Persian civilians starving after years of war and deprivation. In 1918 a small force of 400 British troops under General Dunsterville moved into the Trans-Caucasus from Persia in a bid to encourage local resistance to German and Ottoman armies who were about to invade the Baku oilfields. Although they later withdrew back into Persia, they did succeed in delaying the Turks access to the oil almost until the Armistice. In addition, the expedition's supplies were used to avert a major famine in the region, and a camp for 30,000 displaced refugees was created near the Mesopotamian frontier.

By World War I, Persia was not the world power it had once been. It had become a tool in the political battles of other empires. In 1919, northern Persia was occupied by the British General William Edmund Ironside to enforce the Turkish Armistice conditions and assist General Dunsterville and Colonel Bicherakhov contain Bolshevik influence (of Mirza Kuchak

Khan) in the north. Britain also took tighter control over the increasingly lucrative oil fields.

In 1925, Reza Shah Pahlavi seized power from the Qajars and established the new Pahlavi dynasty, the last Persian monarchy before the establishment of the Islamic Republic. However, Britain and the Soviet Union remained the influential powers in Persia into the early years of the Cold War.

On March 21, 1935, Iran was officially accepted as the new name of the country. After Persian scholars' protests to this decision, in 1953 Mohammad Reza Shah announced both names "Iran" and "Persia" could be used.

## ISLAMIC GOLDEN AGE

During the Islamic Golden Age (fl. 622 AD - 1258 AD in Gregorian calendar; or 1 AH - 656 AH in Islamic calendar), scholars and engineers of the Islamic world contributed enormously to philosophy, science, and technology, both by preserving and building upon earlier traditions and by adding their own inventions and innovations. Scientific and intellectual achievements blossomed in the Golden Age, and passed on to Europe to be expanded upon in the European Renaissance. The Golden Age of Islamic history roughly corresponds to the Middle Ages or Dark Ages in Christian Europe.

### Foundations

During the early Muslim conquests, the Muslim Arabs led by Khalid ibn al-Walid conquered the Sassanid Persian Empire and much of the Byzantine Roman Empire, establishing the Arab Empire across the Middle East, Central Asia, and North Africa, followed by further expansions across Pakistan, southern Italy and the Iberian Peninsula. As a result, the Islamic governments inherited "the knowledge and skills of the ancient Middle East, of Greece, of Persia and of India. They added new and important innovations from outside, such as positional numbering from Ancient India," as Bernard Lewis wrote in What Went Wrong?

Another innovation was paper - originally a secret tightly guarded by the Chinese. The art of papermaking was obtained from two prisoners at the Battle of Talas (751), resulting in paper mills being built in Samarkand and Baghdad. The Arabs improved upon the Chinese techniques using linen rags instead of mulberry bark.

Much of this learning and development can be linked to geography. Even prior to Islam's presence, the city of Mecca served as a center of trade in Arabia and the Islamic prophet Muhammad was a merchant.

The tradition of the pilgrimage to Mecca became a center for exchanging ideas and goods. The influence held by Muslim merchants over African-Arabian and Arabian-Asian trade routes was tremendous. As a result, Islamic civilization grew and expanded on the basis of its merchant economy, in contrast to their Christian, Indian and Chinese peers who built societies from an agricultural landholding nobility.

Merchants brought goods and their faith to China, India (the Indian subcontinent now has over 450 million followers), Southeast Asia (which now has over 230 million followers), and the kingdoms of Western Africa and returned with new inventions. Merchants used their wealth to invest in textiles and plantations.

Aside from traders, Sufi missionaries also played a large role in the spread of Islam, by bringing their message to various regions around the world. The principal locations included: Persia, Ancient Mesopotamia, Central Asia and North Africa. Although, the mystics also had a significant influence in parts of Eastern Africa, Ancient Anatolia (Turkey), South Asia, East Asia and Southeast Asia.

### Arts

The golden age of Islamic (and/or Muslim) art lasted from 750 to the 16th century, when ceramics, glass, metalwork, textiles, illuminated manuscripts, and woodwork flourished. Lusterous glazing became the greatest Islamic contribution to ceramics. Manuscript illumination became an important and greatly respected art, and portrait miniature painting flourished in Persia. Calligraphy, an essential aspect of written Arabic, developed in manuscripts and architectural decoration.

## Philosophy

Only in philosophy were Islamic scholars prevented from putting forth unorthodox ideas. Nevertheless, Arab scientists like al-Kindi, and Ibn Rushd (Averroes) and Persian scientists like Ibn Sina (Avicenna) played a major role in saving the works of Aristotle, whose ideas came to dominate the non-religious thought of the Christian and Muslim worlds. They would also absorb ideas from China, and India, adding to them tremendous knowledge from their own studies. Three speculative thinkers, al-Kindi, al-Farabi, and Ibn Sina (Avicenna), fused Aristotelianism and Neoplatonism with other ideas introduced through Islam.

From Spain the Arabic philosophic literature was translated into Hebrew, Latin, and Ladino, contributing to the development of modern European philosophy. The Jewish philosopher Moses Maimonides, sociologist-historian Ibn Khaldun, Carthage citizen Constantine the African who translated Greek medical texts, and Al-Khwarzimi's collation of mathematical techniques were important figures of the Golden Age.

Ghazali, the famous Persian Jurist and Philosopher, wrote a devastating critique in his Tahafut al-Falasifa on the speculative theological works of Kindi, Farabi and Ibn Sina (Avicenna). Philosophy in the Muslim world never recovered from this critique, even though Ibn Rushd (Averroes) responded strongly in his Tahafut al-Tahafut to many of the points Ghazali raised.

## Sciences

The difference in attitudes of Byzantine scientists and their medieval Muslim peers was firm. Byzantium added little to no new knowledge of science of medicine to the Greco-Roman scientific tradition, stagnating in awe of their classical predecessors. This could perhaps be explained by the fact that the initial Islamic surge out of Arabia had captured three of its most productive cities: Alexandria, Carthage, and Antioch. Because of the loss of a highly skilled and centralized government, as well as continuous and devastating Arab conquests into Anatolia, most Byzantine cities could not support the arts and sciences, and there was a mass return to subsistence farming. The modern scientific method was first developed in the Muslim world, where significant progress in methodology was made, especially in the works of Ibn al-Haytham (Alhazen) in the 11th century.

The most important development of the scientific method was the use of experiments to distinguish between competing scientific theories set within a generally empirical orientation. Ibn al-Haytham (Alhazen) is also known as the "father of optics" for empirically proving that vision occurred because of light rays entering the eye and for using a camera obscura to demonstrate the physical nature of light rays. Ibn al-Haytham has also been described as the "first scientist" for his devlopment of the scientific method.

Most notable Islamic scientists lived and practiced during the Islamic Golden Age. Among the achievements of Muslim scientists and mathematicians during this period included the development of algebra and algorithm (see Muhammad ibn Musa al-Khwarizmi), the invention of spherical trigonometry, the beginning of modern optics and the development of the scientific method (see Ibn al-Haytham), and significant advances in astronomy. These advances included the construction of the first observatory in Baghdad during the reign of Caliph Al-Ma'mun, the collection and correction of previous astronomical data, resolving significant problems in the Ptolemaic model, and improvements of the astrolabe. Several Muslim scientists also considered the possibility of the Earth's rotation on its axis and perhaps a heliocentric solar system.

## Medicine

Medicine was a central part of medieval Islamic culture. Responding to circumstances of time and place, Islamic physicians and scholars developed a large and complex medical literature exploring and synthesizing the theory and practice of medicine. (from the National Library of Medicine digital archives)

Islamic medicine was built on tradition, chiefly the theoretical and practical knowledge developed in Persia, Greece and Rome, and for Islamic scholars. Galen and Hippocrates were pre-eminent authorities, followed by Hellenic scholars in

Alexandria. Islamic scholars translated their voluminous writings from Greek into Arabic and then produced new medical knowledge based on those texts. In order to make the Greek tradition more accessible, understandable, and teachable, Islamic scholars ordered and made more systematic the vast and sometimes inconsistent Greco-Roman medical knowledge by writing encyclopedias and summaries. (from the National Library of Medicine digital archives)

It was through Arabic translations that the West learned of Hellenic medicine, including the works of Galen and Hippocrates. Of equal if not of greater influence in Western Europe were systematic and comprehensive works such as Avicenna's (Ibn Sina) The Canon of Medicine, which were translated into Latin and then disseminated in manuscript and printed form throughout Europe. During the fifteenth and sixteenth centuries alone, The Canon of Medicine was published more than thirty-five times. (from the National Library of Medicine digital archives)

In the medieval Islamic world, hospitals were built in all major cities; in Cairo for example, the Qalawun Hospital could care for 8,000 patients, and a staff that included physicians, pharmacists, and nurses. One could also access a dispensary, and research facility that led to advances in understanding contagious diseases, and research into optics and the mechanisms of the eye. Muslim doctors were removing cataracts with hollow needles over 1000 years before Westerners dared attempt such a task. Hospitals were built not only for the physically sick, but for the mentally sick also. One of the first ever psychiatric hospitals that cared for the mentally ill was built in Cairo.

Medical inventions in the Muslim world included oral anesthesia, inhalant anesthesia, distilled alcohol, medical drugs, chemotherapeutical drugs, injection syringe, and a number of antiseptics and other medical treatments. (See Islamic medicine for more details.)

## Architecture and Engineering

The Great Mosque of Xian in China was completed circa 740, and the Great Mosque of Samarra in Iraq was completed in 847. The Great Mosque of Samarra combined the hypostyle architecture of rows of columns supporting a flat base above which a huge spiraling minaret was constructed.

The Moors began construction of the Great Mosque at Cordoba in 785 marking the beginning of Islamic architecture in Spain and Northern Africa (see Moors). The mosque is noted for its striking interior arches. Moorish architecture reached its peak with the construction of the Alhambra, the magnificent palace/fortress of Granada, with its open and breezy interior spaces adorned in red, blue, and gold. The walls are decorated with stylized foliage motifs, Arabic inscriptions, and arabesque design work, with walls covered in glazed tiles.

The Qol Sharif Mosque of Kazan in Russia was completed in circa 1000. It is still the largest mosque in eastern Europe today.

Another distinctive sub-style is the architecture of the Mughal Empire in India in the 15-17th centuries. Blending Islamic and Hindu elements, the emperor Akbar constructed the royal city of Fatehpur Sikri, located 26 miles (42 km) west of Agra, in the late 1500s and his son Shah Jahan had constructed the mausoleum of Taj Mahal for Mumtaz Mahal in the 1650s.

A significant number of inventions were produced by engineers in the Muslim world. Some of these include the parachute, hang glider, artificial wings, rocket aircraft, water raising machine, cam, brass astrolabe, mechanical astrolabe, camera, pinhole camera, camera obscura, modern chess, coffee, soft drink, fine glass, glasses, glass mirror, cannon, ballistic war machine, counterweight trebuchet, explosive, grenade, gun, firearm, torpedo, iron rocket, rifle, incendiary devices, sulfur bomb, pistol, modern soap, shampoo, kerosene, scribe clock, weight-driven mechanical clock, elephant clock, watch, humanoid robot, segmental gear, mechanical singing bird, kitchen appliances, musical automata, combination lock, hand washing devices, water pump, suction pipe, crankshaft, laminated timber, static balanced wheels, paper models, sand casting, mould box, trick drinking vessels, phlebotomy measures, linkage, hydraulic devices, water level, ewer, movable brass type printing, pendulum, perfumery, trick devices, miswak, homing pigeon, and many other such inventions. (See Islamic inventions for more details.)

### Commerce and Urban Life

From the very beginning, the foundation of Islamic civilization was urban and business oriented, and its growth in population and agriculture is mirrored through its global trade network. Muslim cities grew unregulated, resulting in narrow winding city streets and neighborhoods separated by different ethnic backgrounds and religious affiliations. These qualities proved efficient for transporting goods to and from major commercial centers while preserving the privacy valued by Islamic family life. Suburbs lay just outside the walled city, from wealthy residential communities, to working class semi-slums. City garbage dumps were located far from the city, as were clearly defined cemeteries which were often homes for criminals. A place of prayer was found just near one of the main gates, for religious festivals and public executions. Similarly, Military Training grounds were found near a main gate.

### Urban Life

While varying in appearance due to climate and prior local traditions, Islamic cities were almost always dominated by a merchant middle class. Some peoples' loyalty towards their neighborhood was very strong, reflecting ethnicity and religion, while a sense of citizenship was at times uncommon (but not in every case). The extended family provided the foundation for social programs, business deals, and negotiations with authorities. Part of this economic and social unit were often the tenants of a wealthy landlord.

State power normally focused on Dar al Imara, the governor's office in the citadel. These fortresses towered high above the city built on thousands of years of human settlement. The primary function of the city governor was to provide for defence and to maintain legal order. This system would be responsible for a mixture of autocracy and autonomy within the city. Each neighborhood, and many of the large tenement blocks, elected a representative to deal with urban authorities. These neighborhoods were also expected to organize their young men into a militia providing for protection of their own neighborhoods, and as aid to the professional armies defending the city as a whole.

The head of the family was given the position of authority in his household, although a qadi, or judge was able to negotiate and resolve differences in issues of disagreements within families and between them. The two senior representatives of municipal authority were the qadi and the muhtasib, who held the responsibilities of many issues, including quality of water, maintenance of city streets, containing outbreaks of disease, supervising the markets, and a prompt burial of the dead.

Another aspect of Islamic urban life was waqf, a religious charity directly dealing with the qadi and religious leaders. Through donations, the waqf owned many of the public baths and factories, using the revenue to fund education, and to provide irrigation for Orchards outside the city. Following expansion, this system was introduced into Eastern Europe by Ottoman Turks.

While religious foundations of all faiths were tax exempt in the Muslim world, civilians paid their taxes to the urban authorities, soldiers to the superior officer, and landowners to the state treasury. Taxes were also levied on an unmarried man until he was wed. Instead of zakat, the mandatory charity required of Muslims, non-Muslims were required to pay the jizya, a kind of poll tax.

Animals brought to the city for slaughter were restricted to areas outside the city, as were any other industries seen as unclean. The more valuable a good was, the closer its market was to the center of town. Because of this, booksellers and goldsmiths clustered around the main mosque at the heart of the city.

## Commerce

Guilds were officially unrecognized by the medieval Islamic city, but trades were supervised by an official recognized by the city. Each trade developed its own identity, whose members would attend the same mosque, and serve together in the militia. Slaves were often employed on sugar plantations and salt mines, but more likely as domestic house servants or professional soldiers.

Technology and Industry of Islamic civilization was highly developed. Distillation techniques supported a flourishing perfume industry, while chemical ceramic glazes were developed constantly to compete with ceramics imported from China. A scientific approach to metallurgy made it easier to adopt and improve steel technologies from India and China. Primary exports included manufactured luxuries, such as wood carving, metal and glass, textiles, and ceramics.

The systems of contract relied upon by merchants was very effective. Merchants would buy and sell on commission, with money loaned to them by wealthy investors, or a joint investment of several merchants, who were Muslim, Christian and Jewish. Recently a collection of documents were found in an Egyptian synagogue shedding a very detailed and human light on the life of medieval middle eastern merchants. Business partnerships would be made for many commercial ventures, and bonds of kinship enabled trade networks to form over huge distances. Networks developed during this time enabled a world in which money could be promised by a bank in Baghdad and cashed in Spain, creating the check system of today. Each time items passed through the cities along this extraordinary network, the city imposed a tax, resulting in high prices once reaching the final destination. Regardless, the Muslim world never completely relied on foreign markets, remaining completely self sufficient throughout this period.

Transport was simple yet highly effective. Each city had an area outside its gates where pack animals were assembled, found in the cities markets were large secure warehouses, while accommodations were provided for merchants in cities and along trade routes by a sort of medieval motel.

**Navigation**

Apart from the Nile, Tigris and Euphrates, navigable rivers were uncommon, so transport by sea was very important. Navigational sciences were highly developed making use of a rudimentary sextant known as a kamal to altitudes of stars, and a magnetic compass. When combined with detailed maps of the period, sailors were able to sail across oceans rather than skirt along the coast. Muslim sailors were also responsible for reintroducing large three masted merchant vessels to the Mediterranean.

The caravels used by Italian explorer Christopher Columbus were in fact, based on designs by earlier Muslim Andalusian vessels. An artificial canal linking the Nile with the Gulf of Suez was constructed, conversely linking the Red Sea with the Mediterranean although it silted up several times. A tradition from Muhammad advises his followers to "even travel to China if it means obtaining knowledge." And not long after Muhammad's death, during the Islamic Golden Age, one can observe such travel and exchange with far away lands taking place. For example, Ala'eddin, is honoured in the official history of China's Yuan Dynasty, for having constructed the Counterweight Trebuchet for Kublai Khan.

And we now know that in fact Islam learned paper making from China as a result of this contact, but made the crucial decision to use linen as the raw material for paper, rather than mulberry bark, or other organic matter. The transfer of Chinese technology and the innovation in the use of linen provided a writing material more economical than parchment and more durable than papyrus. It was from Islam that the rest of the world learned to make paper from linen. (from the digital archives of The National Library of Medicine)

**Expeditions Across the Atlantic**

Several contemporary medieval Arabic reports suggest that Muslim explorers from Islamic Spain and Northwest Africa may have travelled in expeditions across the Atlantic Ocean, possibly to the Americas, between the 9th and 14th centuries. Ali al-Masudi (896-856) reported that the navigator Khashkhash Ibn Saeed Ibn Aswad, from Cordoba, Islamic Spain, sailed from Delba (Palos) in 889, crossed the Atlantic, reached an unknown land (which may have been the Americas), and returned with fabulous treasures.

Another Muslim navigator, Ibn Farrukh, from Granada, sailed into the Atlantic on February 999, landed in Gando (Canary islands) visiting King Guanariga, and continued westward where he eventually saw and named two islands, Capraria and Pluitana. He arrived back in Spain in May 999. Other theories suggest that explorers from the Muslim West African Mali Empire may have reached the Americas, or possibly the Hui Chinese Muslim explorer Zheng He according to the 1421 hypothesis. When Christopher Columbus made his first voyage to the Americas in 1492, he was accompanied by a number of Muslim sailors (Andalusian Moors), who travelled with him to the New World.

## Mongol Invasion and the End of the Golden Age

In 1206, Genghis Khan from Central Asia established a powerful Mongol Empire. A Mongolian ambassador to the Caliph in Baghdad is said to have been murdered, which may have been the cause of Hulagu Khan's sack of Baghdad in 1258.

The Mongols conquered most of the Eurasian land mass, including both China in the east and much of the old Islamic caliphate and Islamic Khwarezm, as well as Russia and Eastern Europe in the west, and subsequent invasions of the Levant. Later Mongol leaders, such as Timur, destroyed many cities, slaughtered thousands of people and did irrevocable damage to the ancient irrigation systems of Mesopotamia. These invasions transformed a civil society to a nomadic one.

Eventually, the Mongols that settled in Persia, Central Asia and Russia converted to Islam and in many instances became assimilated into various Muslim Iranian or Turkic peoples (for instance, one of the greatest Muslim astronomers in later times, Ulugh Beg, was a grandson of Timur). The Ottoman Empire rose from the ashes, but the Golden Age was over.

## Causes of Decline

*"The achievements of the Arabic speaking peoples between the ninth and twelfth centuries are so great as to baffle our understanding. The decadence of Islam and of Arabic is almost as puzzling in its speed and completeness as their phenomenal rise. Scholars will forever try to explain it as they try to explain the decadence and fall of Rome. Such questions are*

*exceedingly complex and it is impossible to answer them in a simple way." - George Sarton, The Incubation of Western Culture in the Middle East*

The Islamic civilisation which had at the outset been creative and dynamic in dealing with issues, began to struggle to respond to the challenges and rapid changes it faced during the 12th and 13th century onwards towards the end of the Abbassid rule. Despite a brief respite with the new Ottoman rule, the decline continued until its eventual collapse and subsequent stagnation in the 20th century.

Despite a number of attempts by many writers, historical and modern, none seem to agree on the causes of decline.

The main views on the causes of decline comprise the following: political mismanagement after the early Caliphs in the 8th century, closure of the gates of ijtihad (independent reasoning) in the 12th century and the institutionalisation of taqleed (imitation) rather than ijtihad and creativity, foreign involvement by invading forces and colonial powers (11th, 13th and 19th centuries) and disruption to the cycle of equity (based on Ibn Khaldun's famous model of the rise and fall of civilisations).

Tolerance about different ideas reduced and faded. Seminaries systematically forbade philosophical thought which comprising both natural and theological aspects of world in Islamic context. Even polemic debates were abandoned after the 13th century.

Institutions of science comprising Islamic universities, libraries (including the House of Wisdom), observatories, and hospitals, had been destroyed by foreign invaders like the Mongols and never promoted again. Not only wasn't new publishing equipment accepted but also wide illiteracy overwhelmed Muslim society.

Some historians have recently come to question the traditional picture of decline, pointing to continued astronomical activity as a sign of a continuing and creative scientific tradition through to the 15th century, of which the work of Ibn al-Shatir (1304-1375) in Damascus is considered one of the most noteworthy examples.

## Criticism of Ascribing the Golden Age to Islam

The issue of Islamic Civilization being a misnomer has been raised by a number of recent scholars such as the secular Iranian historian, Dr. Shoja-e-din Shafa in his recent controversial books titled Rebirth and After 1400 Years manifesting the intrinsic contradiction of expressions like "Islamic civilization", "Islamic science", "Islamic medicine", "Islamic astronomy", "Islamic scientists", etc. Shafa states that while religion has been a cardinal foundation for nearly all empires of antiquity to derive their authority from, it does not possess adequate defining factors to advance a kingdom or domain in accumulation and furtherance of science, technology, arts, and culture in a way to justify attribution of such developments to existence and practice of a certain faith within that realm. While various empires in the course of mankind's history advocated and officialized the religion they deemed most appropriate to exercise their absolute authority over the masses, we never ascribe their achievements to the faith they practiced. Ergo, using Islamic attribute for the abovementioned terms is as impertinent as arbitrary concocted namings such as "Christian Civilization" for totality of "Roman Empire" as of Constantine I's reign onwards, "Byzantine Empire" and all subsequent European empires that advocated Christianity one way or another; or "Zoroastrian Architecture" for all the architectural innovations and marvels that pre-Islamic Persian Empire later subsequently loaned to its Muslim conquerers.

But Bernard Lewis opposes this viewpoint.

"There have been many civilizations in human history, almost all of which were local, in the sense that they were defined by a region and an ethnic group. This applied to all the ancient civilizations of the Middle East-Egypt, Babylon, Persia; to the great civilizations of Asia-India, China; and to the civilizations of Pre-Columbian America. There are two exceptions: Christendom and Islam. These are two civilizations defined by religion, in which religion is the primary defining force, not, as in India or China, a secondary aspect among others of an essentially regional and ethnically defined civilization. Here, again, another word of explanation is necessary."

# 8

# Political Geography

Political geography is a field of human geography that is concerned with politics. It is closely related to geopolitics, which is seen as the strategic, military and governmental application of political geographies. It is also closely related to International Relations.

## AREAS OF STUDY

Political geography is interested in the relationship between political power or politics and geography. Geography often influences political decisions and vice versa political power influences geographical space. It is often associated with the study of the practices of sovereign states, but it considers politics of all scales, from those of the United Nations to those of day-to-day life.

In particular, then, modern political geography often considers:

- How and why states are organized into regional groupings, both formally (e.g. the European Union) and informally (e.g. the Third World)
- The relationship between states and former colonies, and how these are propagated over time through neo-colonialism
- The relationship between a government and its people
- The relationships between states including international trades and treaties
- The functions, demarcations and policings of boundaries

- How imagined geographies have political implications
- The influence of political power on geographical space
- The study of election results (election geography)

## HISTORY

### Pre-World War II

The term political geography was first used by Friedrich Ratzel in his book 'Politische Geographie', published in German in 1897. Geopolitics was then coined by the Swede Rudolf Kejell.

The discipline gained attention largely through the work of Sir Halford Mackinder in England and his formulation of the Heartland Theory in 1904. This theory involved concepts diametrically opposed to the notion of Alfred Thayer Mahan about the significance of navies (Mahan coined the term sea power) in world conflict. The Heartland theory, on the other hand, hypothesized the possibility for a huge empire to be brought into existence which didn't need to use coastal or transoceanic transport to supply its military industrial complex, and that this empire could not be defeated by all the rest of the world coalitioned against it.

The Heartland Theory depicted a world divided into a Heartland (Eastern Europe/Western Russia); World Island (Eurasia and Africa); Peripheral Islands (British Isles, Japan, Indonesia and Australia) and New World (The Americas). Mackinder claimed that whoever controlled the Heartland would have control of the world. He used this warning to politically influence events such as the Treaty of Versailles, where buffer states were created between the USSR and Germany, to prevent either of the them controlling the Heartland.

Ratzel, at the same time, was creating a theory on states based around the concepts of Lebensraum and Social Darwinism. This stated that states were 'organisms' that needed sufficient room in which to live. Expansionist inclinations - such as the British of French empires were a result of this. Both these two writers created the idea of a political and geographical science, with an objective God's Eye View of the world.

Pre-World War II political geography was concerned largely with these issues of global power struggles and influencing state policy. The above theories were both taken on board by German geopoliticians (see Geopolitik) such as Karl Haushofer who - they claim inadvertently - greatly influenced Nazi political theory. The politics that were legitimated by 'scientific' theories such as a 'neutral' requirement for state expansion are those that were engaged in during World War Two.

Though modern geographers have been more sympathetic to the likes of Haushofer, suggesting that he and his colleagues actually did believe that they were conducting neutral scientific study, it is also almost impossible that they could not have foreseen how their results would be used; Haushofer, in particular, actually tutored Rudolf Hess and is anecdotely said to have passed a copy of Ratzel's Politische Geographie onto Adolf Hitler, whilst he was writing Mein Kampf.

## Cold War Geopolitics

After World War II, the demonization of Nazi geopolitics lead to a fall in the popularity and legitimacy of the subject. Geographers of this time became more engaged in economic geography, and regional geography. Geography as a discipline went through what is called the quantitative revolution in the 1960s, moving it from a descriptive subject to a more rigorous, theoretically-grounded discipline. Ironically, this reflected closer the stances of inter-war geopoliticians, who cast themselves as scientific observers and analysers.

During this time, when geography took a positivist stance, political geography remained a fairly narrow subject. It was still interested in the division of the world into different groupings - theories such as core and periphery were dominant, with a Marxist informed critique of capitalism.

## Critical Political Geography

Critical political geography is mainly concerned with the criticism of traditional political geographies. Focusing on the expansion of geographical knowledge, within each critical argument is generally nested an alternative cause for the issue at hand, such as within Andrew Feenberg's Critical Theory of Technology.

## GEOPOLITICS

Geopolitics is the study that analyzes geography, history and social science with reference to spatial politics and patterns at various scales (ranging from home, city, region, state to international and cosmopolitics). It examines the political, economic (see geoeconomics) and strategic significance of geography, where geography is defined in terms of the location, size, function, and relationships of places and resources.

The term was coined by Rudolf Kjellén, a Swedish political scientist, at the beginning of the 20th century. Kjellén was inspired by the German geographer Friedrich Ratzel, who published his book Politische Geographie (political geography) in 1897, popularized in English by American diplomat Robert Strausz-Hupé, a faculty member of the University of Pennsylvania.

### Halford Mackinder

The doctrine of Geopolitics gained attention largely through the work of Sir Halford Mackinder in England and his formulation of the Heartland Theory in 1904. The doctrine involved concepts diametrically opposed to the notion of Alfred Thayer Mahan about the significance of navies (he coined the term sea power) in world conflict.

The Heartland theory hypothesized the possibility for a huge empire being brought into existence in the Heartland, which wouldn't need to use coastal or transoceanic transport to supply its military industrial complex but would instead use railways, and that this empire couldn't be defeated by all the rest of the world against it.

The basic notions of Mackinder's doctrine involve considering the geography of the Earth as being divided into two sections, the World Island, comprising Eurasia and Africa; and the Periphery, including the Americas, the British Isles, and Oceania. Not only was the Periphery noticeably smaller than the World Island, it necessarily required much sea transport to function at the technological level of the World Island, which contained sufficient natural resources for a developed economy. Also, the industrial centers of the Periphery were necessarily located in widely-separated locations.

The World Island could send its navy to destroy each one of them in turn. It could locate its own industries in a region further inland than the Periphery could,so they would have a longer struggle reaching them, and would be facing a well-stocked industrial bastion. This region Mackinder termed the Heartland. It essentially comprised Ukraine, Western Russia, and Mitteleuropa.

The Heartland contained the grain reserves of Ukraine, and many other natural resources. Mackinder's notion of geopolitics can be summed up in his saying "Who rules East Europe commands the Heartland. Who rules the Heartland commands the World-Island. Who rules the World-Island commands the world." His doctrine was influential during the World Wars and the Cold War, for Germany and later Russia each made territorial strides toward the Heartland.

Mackinder's geopolitical theory has been criticised as being too sweeping, his interpretation of human history and geography to simple and mechanistic. In his analysis of the importance of mobility, and the move from sea to rail transport, he failed to predict the revolutionary impact of air power. Critically also he underestimated the importance of social organization in the development of power.

## Other Theories

After World War I, Kjellen's thoughts and the term were picked up and extended by a number of scientists: in Germany by Karl Haushofer, Erich Obst, Hermann Lautensach and Otto Maull; in England, Mackinder and Fairgrieve; in France Vidal de la Blache and Vallaux. In 1923 Karl Haushofer founded the "Zeitschrift für Geopolitik" (Journal for Geopolitics), which developed as a propaganda organ for Nazi-Germany. However, more recently Haushofer's influence within the Nazi Party has been questioned (O'Tuathail, 1996) since Haushofer failed to incorporate the Nazis' racial ideology into his work.

Anton Zischka published Afrika, Europas Gemischftaufgabe Tummer (Africa, Complement of Europe) in 1952, where he proposed a kind of North-South Empire, from Stockholm to Johannesburg.

Since then, the word geopolitics has been applied to other theories, most notably the notion of the Clash of Civilizations by Samuel Huntington.

In a peaceable world, neither sea lanes nor surface transport are threatened; hence all countries are effectively close enough to one another physically. It is in the realm of the political ideas, workings, and cultures that there are differences, and the term has shifted more towards this arena, especially in its popular usage. Traditionally, it strictly applies to geography's effect on politics.

**Definitions**

*"In the abstract, geopolitics traditionally indicates the links and causal relationships between political power and geographic space; in concrete terms it is often seen as a body of thought assaying specific strategic prescriptions based on the relative importance of land power and sea power in world history... The geopolitical tradition had some consistent concerns, like the geopolitical correlates of power in world politics, the identification of international core areas, and the relationships between naval and terrestrial capabilities." -Oyvind Osterud, The Uses and Abuses of Geopolitics, Journal of Peace Research, no. 2, 1988, p. 191*

*"By geopolitical, I mean an approach that pays attention to the requirements of equilibrium." Henry Kissinger in Colin S Gray, G R Sloan. Geopolitics, Geography, and Strategy. Portland: Frank Cass Publishers, 1999.*

*"Geopolitics is studying geopolitical systems. The geopolitical system is, in my opinion, the ensemble of relations between the interests of international political actors, interests focused to an area, space, geographical element or ways. - Vladimir Toncea, Geopolitical evolution of borders in Danube Basin, PhD 2006.*

## Historical Geography

Historical geography is the study of the human, physical, fictional, theoretical, and "real" geographies of the past. Historical geography studies a wide variety of issues and topics. A common theme is the study of the geographies of the past and how a place or region changes through time. Many historical geographers study geographical patterns through time, including how people have interacted with their environment, and created the cultural landscape.

Historical geography seeks to determine how cultural features of various societies across the planet emerged and evolved, by understanding their interaction with their local environment and surroundings.

For some in the United States, the term historical geography has a more specialized meaning: the name given by Carl Ortwin Sauer of the University of California, Berkeley to his program of reorganizing cultural geography (some say all geography) along regional lines, beginning in the first decades of the 20th century.

To Sauer, a landscape and the cultures in it could only be understood if all of its influences through history were taken into account: physical, cultural, economic, political, environmental. Sauer stressed regional specialization as the only means of gaining sufficient expertise on regions of the world. Sauer's philosophy was the principal shaper of American geographic thought in the mid-20th century.

Regional specialists remain in academic geography departments to this day. But some geographers feel that it harmed the discipline; that too much effort was spent on data collection and classification, and too little on analysis and explanation. Studies became more and more area-specific as later geographers struggled to find places to make names for themselves. These factors may have led in turn to the 1950s crisis in geography, which raised serious questions about geography as an academic discipline in the United States.

This sub-branch of human geography is closely related to history and environmental history.

## Time Geography

Time geography or time-space geography traces its roots back to the Swedish geographer Torsten Hägerstrand who stressed the temporal factor in the spatial human activities. The time-space path, devised by Hägerstrand, shows the movement of an individual in the spatial-temporal environment with the constraints placed on the individual by these two factors. Three categories of constraints were identified by Hägerstrand:

1. Authority (limits of accessibility to certain places or domains placed on individuals by for example authorities),
2. capability (limitations of movement by individuals. For example, movement is restricted by biological factors, the need for food, drink and sleep.) and
3. coupling (for how long an individual must interact with other individuals in order to finish a task).

The methods associated with time geography have been criticised by a number of postmodern and feminist geographers.

## Regional Geography

Regional geography is a study of regions throughout the world in order to understand or define the unique characteristics of a particular region which consists of natural as well as human elements. Attention is paid also to regionalization which covers the proper techniques of space delineation into regions.

Regional geography is also considered as a certain approach to study in geographical sciences (similar to quantitative geography or bunch of critical geographies). This approach to study was prevailing during the second half of the 19th and the first half of the 20th century also known as a period of prevailing regional geography paradigm when regional geography took the central position in geographical sciences. It was later criticised for its descriptiveness and the lack of theory (regional geography as an empirical approach of geographical sciences). Massive criticism was leveled against this approach in the fifties and during the quantitative revolution. Main critics were Kimble and Schaefer.

Regional geography paradigm has had impact on many geographical sciences (see regional economic geography or regional geomorphology). Todays regional geography is still thought in some universities as study of the major regions of the world, such as Northern and Latin America, Europe, and Asia and their countries. In addition, the notion of a city-regional approach to the study of geography gained some credence in the mid-1990s after works by people such as Saskia Sassen, it was however heavily criticized by Peter Storper.

Notable regional geographers were Alfred Hettner from Germany with his concept of chorology, Vidal de la Blache from France with the possibilism approach (possibilism as a softer notion of environmental determinism) and United States geographer Richard Hartshorne with his areal differenciation concept.

Some geographers have also attempted to reintroduce a certain amount of regionalism since the 1980s. These involve a complex definition of regions and their interactions with other scales.

## Tourism Geography

Tourism Geography or Geotourism is the study of travel and tourism as an industry, as a human activity, and especially as a place-based experience. From a geographical point of view, tourism consists of the places of tourist origin (or tourist generating areas), tourist destinations (or places of tourism supply), and the relationship (connections) between origin and destination places, which includes transportation routes, business relationships, and traveler motivations.

Geography is fundamental to the study of tourism because tourism is geographical in nature. Tourism occurs in places, it involves movement and activities across geographic space (between places), and it is an activity in which both place characteristics and personal self-identities are formed through the relationships that are created among places, landscapes and people.

Physical geography provides the essential background against which tourism places are created, and environmental impacts and concerns are major issues that must be considered in managing the development of tourism places.

Human geography provides an understanding of the social and economic relationships that exist in providing tourism and U.S. National Park Service nature-based outdoor recreation opportunities and activities, as well as the special meaning that these places have to individuals. A destination's sense of place is often a key element behind tourism development -- after all, without the uniqueness and diversity of places, tourism would be mundane and uninteresting.

## Strategic Geography

Strategic geography is concerned with the control of, or access to, spatial areas that have an impact on the security and prosperity of nations. Spatial areas that concern strategic geography change with human needs and development. This field is a subset of human geography, itself a subset of the more general study of geography. It is also related to geostrategy.

## Military Geography

Military geography is an attempt to understand the geopolitical sphere within a military context. OCOKA is a useful acronym in the study of military geography: Obstacles, cover and concealment, observation, key terrain and avenues of approach.

## Feminist Geography

Feminist geography is an approach to study in human geography which applies the theories, methods and critiques of feminism to the study of the human environment, society and geographical space. It is (usually) biased in a feminist way.

## Areas of Study

Feminist geography is often considered part of a broader postmodern approach to the subject which is not primarily concerned with the development of conceptual theory in itself but rather focuses on the real experiences of individuals and groups in their own localities, upon the geographies that they live in within their own communities.

There are a number of strands to feminist geography, which are not completely distinct, and include:

- Geographic differences in gender relations and gender equality
- The geography of women - spatial constraints, welfare geography
- The construction of gender identity through the use and nature of spaces and places
- Geographies of sexuality. (See: Queer theory)
- Childhood and youth studies

In addition to its analysis of the real world, it also critiques existing geographical and social studies, arguing that academic traditions are deliniated by patriarchy, and that contemporary studies which do not confront the nature of previous work reinforce the masculinist bias of academic study.

**Examples:** A simple example would be to consider the way that urban planning has a gender dimension with the expectation that men travel to a distant location for employment while women are involved with child care, basic shopping, and domestic functions in a suburban location.

Why have cities evolved in this way, why have New towns been planned in this way, and by whom and with what ideology?

Another illustration of this approach to geography is to study gender differences in terms of personal access, mobility and safety, especially in respect to the design and use of urban space and open places such as public parks and footpaths.

The experience of walking alone through a city centre late at night or public park even in daytime is likely to be different for males and females and also for people of different ages, cultures and so on.

However, feminist geography is not limited to the local scale. One example of a global topic of feminist geography research is the worldwide migration of women from the Third World to the First World to perform domestic labor and sex work.

## Children's Geographies

Children's geographies is an area of study in human geography, studying the places and spaces of children's lives.

An interest in children's geographies has developed in academic human geography since the beginning of the 1990s, although there were notable studies in the area before that date.

This development emerged from the realisation that previously human geography had largely ignored the everyday lives of children, who (obviously) form a significant section of society, and who have specific needs and capacities, and who may experience the world in very different ways.

Thus children's geographies can in part be seen in parallel to an interest in gender in geography and feminist geography in so much as their starting points were the gender blindness of mainstream academic geography.

Children's geographies rests on the idea that children as a social group share certain characteristics which are experientially, politically and ethically significant and which are worthy of study. The pluralisation in the title is intended to imply that children's lives will be markedly different in differing times and places and in differing circumstances such as gender, family, and class.

Children's geographies is sometimes coupled with, and yet distinguished from the geographies of childhood. The former has an interest in the everyday lives of children; the latter has an interest in how (adult) society conceives of the very idea of childhood and how this impinges on children's lives in many ways. This includes imaginations about the nature of children and the related (spatial) implications.

There are a whole range of focii with children's geographies including children and the city, children and the countryside, children and technology, children and nature, children and globalisation, methodologies of researching children's worlds and the ethics of doing so; see the otherness of childhood.

There is now a journal of Children's Geographies which will give readers a good idea of the growing range of issues, theories and methodologies of this developing and vibrant sub-discipline.

## Children's Culture

Children's culture can be defined in a great number of ways and suffers from being an incredibly broad category. In recent times the study of children's cultural artifacts, children's media and literature and the myths and discourses spun around the notion of childhood have all come under scrutiny within academia, primarily in cultural studies, media studies and literature departments.

## Children's Street Culture

Children's street culture refers to the cumulative culture created by young children. Collectively, this body of knowledge is passed down from one generation of urban children to the next, and can also be passed between different groups of children (e.g. in the form of crazes, such as making guys for Bonfire Night - see Beck 1984). It is often strongest in urban working class industrial districts where children are traditionally free to "play out" (thus creating Children's street culture) in the streets for long periods without supervision. It is most common in children between the ages of seven and twelve.

It is not to be confused with the commercial narrative media-culture produced for children (e.g., comics, television, mass-produced toys, and clothing), although it may overlap. Children's street culture is invented and largely sustained by children themselves, although it may come to incorporate fragments of media culture and toys in its activities. It can also incorporate scavenged materials such as old car seats, tyres, planks, bricks, etc.

Although it varies from place to place, research shows that it appears to share many commonalities across many cultures. It is a traditional phenomenon that has been closely investigated and documented during the 20th century by anthropologists and folklorists such as Iona Opie; street photographers such as Roger Mayne, Helen Levitt, David Trainer, Humprey Spender and Robert Doisneau; urbanists such as Colin Ward and Robin Moore, as well as being described in countless novels of childhood. It has occasionally been central to feature films, such as the Our Gang ("Little Rascals") series, Ealing's Hue & Cry (1947) and some Children's Film Foundation films such as Go Kart, Go! and Soap Box Derby.

Since the widespread use of the car, children's street culture has often been forced to retreat to pavements and backstreets, and then into parks and playgrounds.

Since the advent of distractions such as video games, television and peer pressure concerns have been expressed about the vitality or even the survival of children's own street culture for it began to die. Since the mid 1990s in some nations, parental fears about pedophiles have led many adults to forbid unsupervised outdoor play.

In 1997, The Miami New Times published Lynda Edwards' "Myths Over Miami", which describes a huge consistent mythology spreading among young homeless children in the American South. There is no known verification or confirmation that the mythology she describes actually exists, but these "secret stories" are clearly based on known elements of street culture, such as labeling certain places "haunted" or recycling legends of dangerous spirits such as Mary Worth.

The article was the basis for Mercedes Lackey's novel Mad Maudlin, co-written with Rosemary Edghill.

## Environmental Geography

Environmental geography is the branch of geography that describes the spatial aspects of interactions between humans and the natural world. It requires an understanding of the traditional aspects of physical and human geography, as well as the ways in which human societies conceptualize the environment.

Environmental geography has emerged as a bridge between human and physical geography as a result of the increasing specialisation of the two sub-fields. Furthermore, as human relationship with the environment has changed as a result of globalisation and technological change a new approach was needed to understand the changing and dynamic relationship. Examples of areas of research in environmental geography include disaster management, environmental management, sustainability and political ecology.

## Geomatics

Geomatics is a branch of geography that has emerged since the quantitative revolution in geography in the mid 1950s. Geomatics involves the use of traditional spatial techniques used in cartography and topography and their application to computers. Geomatics has become a widespread field with many other disciplines using techniques such as GIS and remote sensing. Geomatics has also lead to a revitalisation of some geography departments especially in Northern America where the subject had a declining status during the 1950s.

Geomatics encompasses a large area of fields involved with spatial analysis, such as Cartography, Geographic information systems (GIS), Remote sensing and GPS.

## Regional Geography

Regional geography is a branch of geography that studies the regions of all sizes across the Earth. It has a prevailing descriptive character. The main aim is to understand or define the uniqueness or character of a particular region which consists of natural as well as human elements. Attention is paid also to regionalization which covers the proper techniques of space delimitation into regions.

Regional geography is also considered as a certain approach to study in geographical sciences (similar to quantitative or critical geographies, for more information see History of geography).

## Related Fields

- Urban planning, regional planning and spatial planning: use the science of geography to assist in determining how to develop (or not develop) the land to meet particular criteria, such as safety, beauty, economic opportunities, the preservation of the built or natural heritage, and so on. The planning of towns, cities and rural areas may be seen as applied geography.

- Regional science: In the 1950s the regional science movement led by Walter Isard arose, to provide a more quantitative and analytical base to geographical questions, in contrast to the descriptive tendencies of traditional geography programs. Regional science comprises the body of knowledge in which the spatial dimension plays a fundamental role, such as regional economics, resource management, location theory, urban and regional planning, transport and communication, human geography, population distribution, landscape ecology, and environmental quality.
- Interplanetary Sciences: While the discipline of geography is normally concerned with the Earth, the term can also be informally used to describe the study of other worlds, such as the planets of the solar system, and even beyond. The study of systems larger than the earth itself usually forms part of Astronomy or Cosmology. The study of other planets is usually called planetology. Alternative terms such as areology (the study of Mars) have been proposed but are not widely used.

## GEOGRAPHICAL TECHNIQUES

As spatial interrelationships are key to this synoptic science, maps are a key tool.

Classical cartography has been joined by a more modern approach to geographical analysis, computer-based geographic information systems (GIS).

In their study, geographers use four interrelated approaches:

- Systematic - Groups geographical knowledge into categories that can be explored globally.
- Regional - Examines systematic relationships between categories for a specific region or location on the planet.
- Descriptive - Simply specifies the locations of features and populations.
- Analytical - Asks why we find features and populations in a specific geographic area.

### Cartography

Cartography studies the representation of the Earth's surface with abstract symbols (map making). Although other subdisciplines of geography rely on maps for presenting their analyses, the actual making of maps is abstract enough to be regarded separately. Cartography has grown from a collection of drafting techniques into an actual science.

Cartographers must learn cognitive psychology and ergonomics to understand which symbols convey information about the Earth most effectively, and behavioral psychology to induce the readers of their maps to act on the information.

They must learn geodesy and fairly advanced mathematics to understand how the shape of the Earth affects the distortion of map symbols projected onto a flat surface for viewing. It can be said, without much controversy, that cartography is the seed from which the larger field of geography grew. Most geographers will cite a childhood fascination with maps as an early sign they would end up in the field.

### Geographic Information Systems

Geographic information systems (GIS) deal with the storage of information about the Earth for automatic retrieval by a computer, in an accurate manner appropriate to the information's purpose.

In addition to all of the other subdisciplines of geography, GIS specialists must understand computer science and database systems. GIS has revolutionized the field of cartography; nearly all mapmaking is now done with the assistance of some form of GIS software. GIS also refers to the science of using GIS software and GIS techniques to represent, analyze and predict spatial relationships. In this context, GIS stands for Geographic Information Science.

### Remote Sensing

Remote sensing can be defined as the art and science of obtaining information about Earth features from measurements made at a distance. Remotely sensed data comes in many forms such as satellite imagery, aerial photography and data obtained from hand-held sensors. Geographers increasingly use remotely sensed data to obtain information about the Earth's land surface, ocean and atmosphere because it:

a) supplies objective information at a variety of spatial scales (local to global),

b) provides a synoptic view of the area of interest,

c) allows access to distant and/or inaccessible sites,

d) provides spectral information outside the visible portion of the electromagnetic spectrum, and

e) facilitates studies of how features/areas change over time.

Remotely sensed data may be analyzed either independently of, or in conjunction with, other digital data layers (e.g., in a Geographic Information System).

## Geographic Quantitative Methods

Geostatistics deal with quantitative data analysis, specifically the application of statistical methodology to the exploration of geographic phenomena. Geostatistics is used extensively in a variety of fields including: hydrology, geology, petroleum exploration, weather analysis, urban planning, logistics, and epidemiology.

The mathematical basis for geostatistics derives from cluster analysis, discriminant analysis, and non-parametric statistical tests, and a variety of other subjects. Applications of geostatistics rely heavily on Geographic Information Systems, particularly for the interpolation (estimate) of unmeasured points. Geographers are making notable contributions to the method of quantitative techniques.

Geographic qualitative methods, or ethnographical; research techniques, are used by human geographers. In cultural geography there is a tradition of employing qualitative research techniques also used in anthropology and sociology. Participant observation and in-depth interviews provide human geographers with qualitative data.

## SELECTIVE LIST OF NOTABLE GEOGRAPHERS

- **Eratosthenes (276BC - 194BC)** - calculated the size of the Earth.
- **Ptolemy (c.90–c.168)** - compiled Greek and Roman knowledge into the book Geographia.
- **Gerardus Mercator (1512-1594)** - innovative cartographer produced the mercator projection
- **Alexander Von Humboldt (1769–1859)** - Considered Father of modern geography, published the Kosmos and founder of the sub-field biogeography.
- **Carl Ritter (1779-1859)** - Considered Father of modern geography. Occupied the first chair of geography at Berlin University.
- **Arnold Henry Guyot (1807-1884)** - noted the structure of glaciers and advanced understanding in glacier motion, especially in fast ice flow.
- **William Morris Davis (1850-1934)** - father of American geography and developer of the cycle of erosion.
- **Paul Vidal de la Blache (1845-1918)** - founder of the French school of geopolitics and wrote the principles of human geography.
- **Sir Halford John Mackinder (1861-1947)** - Co-founder of the LSE, Geographical Association of which he later became president, Reading University and author of The Geographical Pivot of History and Heartland Theory.
- **Walter Christaller (1893-1969)** - human geographer and inventor of Central Place Theory.
- **Yi-Fu Tuan (1930-)** - Chinese-American scholar credited with starting Human Geography as a discipline.
- **David Harvey (1935-)** - Marxist geographer and author of theories on spatial and urban geography.
- **Michael Frank Goodchild (1944-)** - prominent GIS scholar and winner of the RGS founder's medal in 2003.
- **Nigel Thrift (1949-)** - originator of non-representational theory.

## GEOGRAPHY HALL OF FAME

- **Largest Continent:** Asia, 17,212,000 square miles
- **Smallest Continent:** Australia, 3,132,000 square miles
- **Highest Mountain:** Mount Everest, Himalayan Mountains, Nepal-Tibet, 29,035 feet above sea level
- **Lowest Point on Land:** The Dead Sea, Israel-Jordan, water surface 1,349 feet below sea level
- **Deepest Underwater Trench:** Mariana Trench, 200 miles southwest of Guam in the Pacific Ocean, 36,198 feet below the ocean surface
- **Largest Sea:** The Mediterranean Sea, 1,144,800 square miles
- **Highest Lake:** The highest navigable lake is Lake Titicaca in Peru, 12,500 feet above sea level
- **Lowest Lake:** The Dead Sea, Israel-Jordan, surface of water 1,349 feet below sea level
- **Largest Lake:** Caspian Sea, 152,239 square miles
- **Largest Freshwater Lake:** Lake Superior, U.S.-Canada, 31,820 square miles
- **Deepest Ocean:** Pacific Ocean, average depth 13,215 feet
- **Largest Ocean:** Pacific Ocean, 60,060,700 square miles
- **Smallest Ocean:** Arctic Ocean, 5,427,000 square miles
- **Largest Gulf:** Gulf of Mexico, 615,000 square miles
- **Largest Bay:** The Bay of Bengal, 1,300,000 square miles
- **Largest Island:** Greenland, 839,999 square miles
- **Largest Peninsula:** Arabia, 1,250,000 square miles
- **Largest Archipelago:** Indonesia, 3,500-mile stretch of 17,000 islands
- **Largest Gorge:** Grand Canyon, Colorado River, Arizona, U.S., 217 miles long, 4–18 miles wide, 1 mile deep

- **Deepest Gorge:** Hells Canyon, Snake River, Idaho, 7,900 feet deep
- **Longest Mountain Range:** The Andes of South America, 5,000 miles
- **Longest River:** The Nile, Africa, 4,180 miles
- **Shortest River:** The Roe, Montana, U.S., 200 feet long
- **Largest River:** The Amazon, South America, basin of 2,500,000 square miles
- **Longest Estuary:** Ob River, Russia, 550 miles long, up to 50 miles wide
- **Largest Lagoon:** Lagoa dos Patos, Brazil, 150 miles long, 4,500 square miles
- **Largest Waterfall:** Angel Falls, Venezuela, 3,212 feet high.

# BIBLIOGRAPHY


Atkins PJ: *The Maltese Food System and the Mediterranean*, Europe, Malta, 1997.

Bernus E: *Coping with Uncertainity in Food Supply*, Oxford University Press, 1988.

Bestor TC: *Theorizing the City: The New Urban Anthropology Reader*, Rutgers University Press, 1999.

Cohen MN: *Food and Evolution*, Temple University Press, 1987.

David, W.: *The Art of Navigation in England in Elizabethan and Early Stuart Times*, London, Hollis and Carter, 1958.

Driver, F.: *Power and Pauperism: the Workhouse System 1834-1884*, Cambridge, Cambridge University Press, 1993.

Duncan, N. : *Destabilizing Geographies of Gender and Sexuality*, London, Routledge, 1995.

Edward, W.: *Postmodern Geographies: The Reassertion of Space in Critical Social Theory*, New York: Verso, 1989.

Fuss, D.: *Essentially Speaking. Feminism, Nature and Difference*, Routledge, London, 1989.

Giddens, A.: *The Transformation of Intimacy: Sexuality, Love and Eroticism in Modern Societies*, Polity Press, Cambridge, 1992.

Hayden, D.: *The Grand Domestic Revolution: A History of Feminist Designs for American Homes, Neighborhoods, and Cities*, MIT Press, Cambridge, 1982.

Jackson, P. and J. Penrose: *Constructions of Race, Place and Nation*, University College London Press, London, 1993.

Janice R. Welsch: *Multiple Voices in Feminist Film Criticism*, Minneapolis, University of Minnesota Press, 1994.

Longhurst, R.: Reflections on and a Vision for Feminist Geography, New Zealand Geographer, 1994.

Lynam, Edward: *The Mapmaker's Art: Essays on the History of Maps*, London, Batchworth, 1953.

MacCarthy-Morrogh, Michael: *The Munster Plantation: English Migration to Southern Ireland 1583-1641*, Oxford, Clarendon, 1986.

Massey, D. B.: *Space, Place and Gender*, The LSE Magazine, 1992.

Monk, J.: *Full Circles: Geographies of Women over the Life Course*, Routledge, London and New York, 1993.

Mundy, Barbara: *The Mapping of New Spain: Indigenous Cartography and the Maps of the Relaciones Geograficas*, Chicago, U of Chicago P, 1996.

Pagden, Anthony: *The Fall of Natural Man. The American Indian and the Origins of Comparative Ethnology*, Cambridge, Cambridge UP, 1982.

Parks, George Bruner: *Richard Hakluyt and the English Voyages*, New York, American Geographical Society, 1928.

Rose, Gillian: *Feminism and Geography: The Limits of Geographical Knowledge*, Cambridge, Polity, 1993.

Schama, Simon: *Landscape and Memory*, London, HarperCollins, 1995.

Shirley, Rodney: *The Mapping of the World: Early Printed World Maps 1472-1700*, London, New Holland, 1983.

Thirsk, Joan: *The Agrarian History of England and Wales*, Cambridge, Cambridge UP, 1967.

Tooley, R. V. and Charles Bricker: *Landmarks of Mapmaking: An Illustrated Survey of Maps and Mapmaking*, Ware, Wordsworth, 1976.

Tyacke, Sarah: *English Map Making 1500-1650: Historical Essays*, London, British Library, 1983.

Valentine, G.: *Mapping Desire: Geographies of Sexualities*, London, Routledge, 1995.

Wallis, Helen: *Historians' Guide to Early British Maps*, London, Royal Historical Society, 1994.

Wood, Denis: *The Power of Maps*, New York, Guilford, 1992.

Yates, Frances: *Theatre of the World*, London, Routledge and Kegan Paul, 1969.

# INDEX

❑❑❑